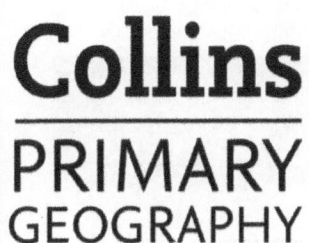

Our planet
Teacher's Guide 2

Stephen Scoffham | Colin Bridge

Geography in the primary school	2
Collins Primary Geography overview	3
Lesson summary	4
Lesson planning	6
Studying the local area	7
Differentiation and progression	8
Assessment	9
Ofsted and the National Curriculum in England	10
Support and guidance	11
Unit-by-unit notes	12
Photocopiable resource matrix	26
Photocopiable resources	30
Geography in the National Curriculum in England	60
World maps	62

Geography in the primary school

Geography is the study of the Earth's surface. It helps children understand the human and physical forces which shape the environment and the way it is changing. Children are naturally interested in their immediate surroundings. They also want to know about places beyond their direct experience. Geography is uniquely placed to satisfy this curiosity.

Geographical enquiries

Geography is an enquiry-led subject that explores fundamental questions such as:

- Where is this place?
- What is this place like (and why)?
- How and why is it changing?
- How does this place compare with other places?
- How and why are places connected?

These questions involve not only finding out about the natural processes which have shaped our environment; they also involve finding out how people have responded to them. Studying this interaction at a range of scales from the local to the global and asking questions about what is happening in the world around us lies at the heart of both academic and school geography.

Geographical perspectives

Geographical perspectives offer a uniquely powerful way of seeing the world. Since the time of the Ancient Greeks, geographers have been attempting to chronicle and interpret their surroundings. One way of seeing links and connections is to think in terms of key concepts and ideas. Three concepts, which have proved particularly useful in a range of settings, are place, space and scale.

- Place focuses attention on specific places (real and imagined) and highlights their character, current activities, changes and development.
- Space focuses attention on the relationship between features and places, and refers to where they are located, the patterns they form and networks connecting them.
- Scale enables geographers to look at the world from very small local sites to international regions.

A layer of additional concepts provides a further way of enhancing geographical understanding. These concepts include pattern, change, movement, interconnections, culture, power, sustainability and environmental impact. Taken in combination, these concepts act as a 'lens' for describing and analysing the complexities of the world around us.

As they conduct their enquiries and investigations geographers make use of some subject-specific skills. Foremost among these are mapwork and the ability to represent spatial information. Geographers also champion the use of digital data which enables them to portray changes and explore different scenarios. The use of maps, charts, diagrams, tables, sketches and other cartographic techniques – all of which allow us to visualise and better understand data – come under the more general heading of 'graphicacy'. Graphicacy is sometimes seen as a key human attribute and a distinguishing feature of geographical thinking.

Geography in primary schools gives children from the earliest ages a fascinating window onto the world. It embraces major concerns such as climate change, migration and biodiversity loss. The challenge for educators is to find ways of providing experiences and selecting content that will help children develop an increasingly deep understanding of the world around them.

Collins Primary Geography overview

Collins Primary Geography is a complete programme for pupils in the primary school and can be used as a structure for teaching geography from ages 5–11 and beyond. At its core are six Pupil Books, each of which has a linked Workbook. This Teacher's Guide provides teaching notes and photocopiable resources for each lesson. Editable Word, PDF and PowerPoint files are available to help you adapt the resources to the needs of your class. Audio files are also available for the stories in Pupil Books 1 and 2.

Aims

The overall aim of the programme is to inspire children with an enthusiasm for geography and to empower them as learners. The underlying principles include a commitment to international understanding in a more equitable world; a concern for the future welfare of the planet; and a recognition that creativity, hope and optimism play a fundamental role in lasting learning. Three different dimensions – connecting to the environment, connecting to each other and connecting to ourselves – are explored throughout the programme in different contexts and at a range of scales. We believe that learning to think geographically in the broadest meaning of the term will help to prepare children for the future and whatever it may hold.

Structure

Collins Primary Geography provides full coverage of the National Curriculum in England framework. Each Pupil Book covers a balanced range of themes and topics and includes case studies with a more precise focus:

- Book 1 *World around me* introduces children to the world at a local scale.
- Book 2 *Our planet* explores the world at a global scale.
- Book 3 *Investigation* encourages pupils to conduct their own research and enquiries.
- Book 4 *Movement* considers how movement affects the physical and human environment.
- Book 5 *Change* includes case studies on how places alter and develop.
- Book 6 *Issues* considers more complex ideas to do with the environment and sustainability.

Although the books are not limited to a specific age group, Book 1 will be particularly suitable for children at the beginning of their formal education. Book 2 is suitable for ages 6–7, Book 3 for ages 7–8, Book 4 for ages 8–9, Book 5 for ages 9–10 and Book 6 for ages 10–11, or children at the end of primary school.

The programme is structured in such a way that key themes are revisited, making it possible to investigate a specific topic in greater depth if required.

Investigations

Enquiries and investigations are an important part of pupils' work in primary geography. Asking questions and searching for answers can help children develop core knowledge, understanding and skills. Fieldwork is time-consuming when it involves travelling to distant locations, but local area work can be equally effective. Many of the exercises in *Collins Primary Geography* focus on the classroom, school building and local environment. We believe that such activities can have a seminal role in promoting long-term positive attitudes towards sustainability and the environment.

Lesson summary

Lesson	Key idea	Geographical vocabulary
Earth in space		
Introduction	Geography is about the world we live in.	
1 Earth, sun and moon	The sun and the moon move through the sky above the Earth.	Earth, sun, moon, sky, tide
2 The planets	The Earth is one of eight planets which go round the sun.	rock, space, planet, journey
3 Day and night	As the Earth spins in space, we get day and night.	night, stars, daytime, pole, equator
4 Land and water	Land and water cover the Earth's surface.	water, land, sea, soil
Planet Earth		
5 A living planet	Water brings life to the Earth. All living things need water.	world, seed, sand, desert, cloud
6 The shape of the land	The land consists of mountains, hills and lowlands.	mountain, river, pool, downstream, current, cliff, waterfall, valley
7 Volcanoes	Volcanoes bring hot rocks to the surface from deep underground.	volcano, erupt, surface, core
8 World wonders	There are many special sights to see in the world.	coast, iceberg, cave, northern lights, beach
Weather and seasons		
9 Experiencing the weather	There are many different types of weather.	weather, rain, wind, shower, gale, snow
10 Different types of weather	We can describe the weather using words and symbols.	symbols, warm, sun, cloud, rain, shower, fog, mist, thunder, lightning
11 Extreme weather	Sometimes the weather is wild and exciting.	field, tornado, hurricane, flood, sandstorm
12 The seasons	In some countries, there is a pattern of seasons during the year.	seasons, pattern, year, spring, summer, autumn, winter
13 Going round the sun	The seasons change as the Earth goes round the sun.	high, low, tilt, towards, away
Local areas		
14 Shelter	Homes give us warmth and shelter.	shelter, houses, camping, tent, item, kit
15 Houses around the world	People build houses in lots of different ways.	brick, wood, stone, concrete, country, city
16 Living in a village	People live together in groups or communities.	village, school, shop, garage, cottage, pond
17 Exploring local streets	There are lots of items in a street which help people live their lives.	street, rubbish bin, lamp post, drain

| 18 Under your feet | The pipes and wires under the pavement provide the things we need in our daily lives. | pipes, wires, pavement, gas, electricity, ground |

Maps and plans

19 Maps and stories	Picture maps can show us about the places in songs and stories.	map, place, hill, bank
20 Treasure island	We can use maps to show places both real or imagined.	town, lighthouse, castle, museum, directions, island, grid
21 Different plans	Plans show the shape of places around us.	plan, shape, room, office, size, semi-detached, terraced
22 The view from above	Plans show what places look like from above.	view, above, rectangle, photograph

The United Kingdom

| 23 Countries and capitals in the United Kingdom | There are four countries in the United Kingdom. Each country has a capital city. | country, England, Northern Ireland, Scotland, Wales, capital city, London, Belfast, Edinburgh, Cardiff |
| 24 Mountains, rivers and seas in the United Kingdom | The United Kingdom has mountains, rivers and seas. | Yr Wyddfa (Snowdon), River Severn, Norfolk Broads |

Different environments

25 Living in the Arctic	The Arctic surrounds the North Pole. It is very cold and snowy.	Arctic, North Pole, walrus, reindeer, polar bear
26 Living in the rainforest	The rainforest lies on the Equator. It is hot and wet.	rainforest, vegetation, parrot, butterfly, jaguar, hummingbird, anteater
27 Living in the desert	Most deserts are very hot and very dry.	shade, kangaroo, lizard, eucalyptus
28 Animals around the world	We share our world with many different plants and creatures.	Greenland, Amazon, China, India, Africa

World maps

| 29 World continents and oceans | The world is divided into continents and oceans. | continent, ocean |
| 30 World countries | There are about 200 countries in the world. | world, countries |

Oracy and critical thinking

Each lesson provides an opportunity for pupils to engage with and explore information and ideas through discussion. 'Talking' panels present questions requiring critical thinking or personal input appropriate for the age group. These activities give all pupils the opportunity to practise speaking with confidence, explaining their ideas, listening and responding to others, and participating in group discussions. You, as the teacher, can facilitate whole class, group or one-on-one discussions by modelling speaking and asking prompting questions. There are also opportunities for pair and groupwork in the other activities. These will further encourage development of oracy and critical thinking.

Lesson planning

Collins Primary Geography has been designed to support both whole-school and individual lesson planning. As you devise your schemes and work out lesson plans, you may find it helpful to ask the following questions. Have you:

- Given children a range of entry points which will engage their enthusiasm and capture their imagination?
- Used a range of teaching strategies which cater for pupils who learn in different ways?
- Thought about how you will draw pupils into class discussion, supporting them to develop their understanding and ideas?
- Thought about using practical activities and games?
- Explored the ways that stories or personal accounts might be integrated with the topic?
- Considered the opportunities for fieldwork?
- Encouraged pupils to use and make maps and diagrams?
- Included examples from around the world to enhance global awareness?
- Questioned whether you are challenging rather than reinforcing stereotypes?
- Checked on links to suitable websites, particularly with respect to research?
- Made use of digital programs to record findings or analyse information?
- Made links to other subjects where there is a natural overlap?
- Promoted geography alongside oracy and literacy skills especially in talking and writing?
- Taken advantage of the opportunities for presentations and class displays?
- Ensured that the pupils are developing geographical skills and meaningful subject knowledge?
- Clarified the knowledge, skills and concepts that will underpin the lesson?
- Identified appropriate learning outcomes or given pupils the opportunity to identify their own ones?

These questions are offered as prompts which may help you to generate stimulating and lively lessons. There is clear evidence that when geography is fun and pupils enjoy what they are doing it can lead to lasting learning. Striking a balance between light-hearted delivery and serious intent is part of the craft of being a teacher. *Always remember to follow the latest advice for practising online safety in research activities.*

Layout of the units

Book 2 has eight units divided into two to five lessons. Each lesson then follows a consistent layout:

Lesson title
Identifies the theme of the lesson. The supporting Workbook unit and photocopiable resource use the same title which makes them easy to identify.

Introductory and summary text
Identifies the theme of the lesson and its geographical significance. This serves both as a learning objective and an assessment statement. (The unit-by-unit notes provide further topic information to give you, as the teacher, additional context for teaching each lesson. These notes are not designed for the pupils).

Story
Introduces the main theme and vocabulary. Use the question to confirm comprehension of the key ideas.

Read the words panel
Highlights the lesson's key geographical words. Children could choose a word and make a picture or a model, and support this with a text label. Or you might use discussion to reinforce understanding.

Amazing fact!
Provides extra information to engage children. Pupils might research additional facts for themselves, use them in a quiz, or add them to a display.

Talking and For you to do panels
Provide ideas for discussion and practical activities. These show how the lesson's learning is significant in children's own lives and the natural environment. (The unit-by-unit notes provide 'Further activities'. These offer additional ideas to develop the lesson theme and opportunities for cross-curricular links).

Studying the local area

The local area is the immediate vicinity around the school and the home. It consists of three different components: the school building, the school grounds, and local streets and buildings. By studying their local area, children will learn about the different features which make their environment distinctive and how it attains a specific character. When they are familiar with their own area, they will then be able to make meaningful comparisons with more distant places.

There are many opportunities to support the lessons outlined in *Collins Primary Geography* with practical local area work. First-hand experience is fundamental to good practice in geography teaching, is a clear requirement in the programme of study and has been highlighted in guidance to Ofsted inspectors. The local area can be used not only to develop ideas from human geography but also to illustrate physical and environmental themes. The checklist below illustrates some of the features which could be identified and studied.

Physical geography

Hill, valley, cliff, mountain, rock, slope, soil, forest

River, stream, pond, lake, estuary, ocean, sea, beach, coast

Slopes, rock, soil, vegetation and other small-scale features

Local weather, seasons and site conditions

Human geography

Origins of settlements (city, town, village), land use (farms) and economic activity

House, cottage, terrace, flat, housing estate

Roads, stations, harbours, ports

Shops, factories and offices

Fire, police, ambulance, health services

Library, museum, park, leisure centre

All work in the local area involves collecting and analysing information. An important way in which this can be achieved is through the use of maps and plans. Other techniques include annotated drawings, bar charts, tables and reports. There will also be opportunities for the children to make presentations in class and perhaps to the rest of the school in assemblies.

Misconceptions

There is a growing body of research which helps practitioners to understand more about how children learn primary geography and the barriers and challenges that they commonly encounter. The way that young children assume that the physical environment was created by people was first highlighted by Jean Piaget. The importance and significance of early childhood misconceptions was further illuminated by Howard Gardner. More recent research has considered how children develop their understanding of maps and places. Children's ideas about other countries and their attitudes to other nationalities form another very important line of enquiry. So, too, do their ideas about climate change and the environment. Some key readings are listed in the references on page 11.

Differentiation and progression

Collins Primary Geography sets out to provide access to the curriculum for children of all abilities. It is structured so that children can respond to and use the material in a variety of ways.

- Each lesson contains a range of stimulus material designed to engage children imaginatively. This means activities can be selected which are appropriate to individual circumstances.
- Considerable emphasis is placed on discussion and critical thinking appropriate for the age group, which allows teachers to frame discussions and respond to pupils according to their level of understanding.
- The Workbooks provide further opportunities to develop and evidence understanding.
- The photocopiable resources allow you to assess understanding of key concepts during the unit.

Teachers will be able to select what they think will be appropriate from a range of resources. There is no need to work through all the material.

Differentiation by outcome

Each lesson starts with a brief introductory text which summarises a key idea in a short paragraph. This is followed by a story, featuring a cast of recurring characters, which explores the main theme in a light-hearted but purposeful manner. Some children will be able to read these stories independently. Others may require support. You may choose to read these stories as a class, as a group or individually, depending on the context of your class.
A comprehension question is given to confirm understanding of the key ideas. Linked discussion questions explore the ideas further, introduce key vocabulary and provide an opportunity for structured talking. Lessons end with a 'For you to do' activity to draw attention to how the lesson relates to the child on a personal level. To extend, some children will be able to consider the underlying geographical concepts. The pace and range of the discussion can be controlled to suit the needs of the class or group.

Differentiation by task

Teachers may decide to complete some of the tasks as class exercises or help learners who require more support by working through the first part of an exercise with them. Classroom assistants could also work with individual children or small groups. Some children could be given extension tasks. Ideas and suggestions for extending each lesson are provided in the information on individual themes (pages 12–25). The Workbooks include a wider range of activities with varying levels of scaffolding to support Pupil Book investigations when required.

Differentiation by process

Children of all abilities benefit from exploring their environment and experiencing the world around them at first hand. Work in the local area also helps to overcome the problems of written communication by focusing on concrete events. There are also opportunities for taking photographs as well as developing geographical vocabulary through observation and discussion.

Finding time for geography

The pressures on the school timetable and the demands of the core subjects make it hard to secure adequate time for primary geography. However, finding ways of integrating geography with mathematics and literacy can be a creative way of increasing opportunities. Geography also has a natural place in a wide range of social studies and current affairs whether local or global. It can be developed through class assemblies and extra-curricular studies. Those who are committed to thinking geographically find a surprising number of ways of developing the subject whatever the accountability regime in which they operate.

Assessment

Assessment is often seen as having two different dimensions.

- *Formative assessment* is an on-going process which provides both pupils and teachers with information about the progress they are making in a piece of work.
- *Summative assessment* occurs at defined points in a child's learning and seeks to establish what they have learnt and how they are performing in relation both to their peers and to nationally agreed standards.

Collins Primary Geography provides opportunities for both formative and summative assessment.

Formative assessment

- The Talking questions invite pupils to discuss a topic, relate it to their previous experience and consider any issues which may arise, thereby yielding information about their current knowledge and understanding.
- The Pupil Book activities and exercises will help pupils develop their geographical knowledge and understanding at a manner which is appropriate to their current level of ability
- The Workbook activities provide an opportunity for pupils to apply their understanding and evidence learning.

Summative assessment

- Each lesson includes a Talking panel which is designed to promote discussion. The questions provided here could be used summatively if required.
- The photocopiable resources (see pages 30–59) can be used to provide further evidence of key concepts, and to track progress of knowledge and skills. Whether used formatively or summatively, they are intended to consolidate understanding and identify gaps in learning to inform future teaching.

Reporting to parents and guardians

Collins Primary Geography is structured around geographical skills, themes and places. As children work through the lessons, they can build up a folder of work and progress through the Workbook. This will provide evidence of mapwork and other practical activities both inside and outside the classroom and provide a rounded portrait of pupil achievement. This will also be a useful resource when teachers report to parents and guardians and show if a child requires further support in geography.

National curriculum reporting

There is a single attainment target for geography and other National Curriculum subjects in the National Curriculum for England. This simply states that

> 'By the end of each key stage, pupils are expected to know, apply and understand the matters, skills and processes specified in the relevant programme of study.'

This means that assessment need not be an onerous burden, and that evidence of pupils' achievement can be built up over a set of years (or a Key Stage). The assessment process can also inform lesson planning.

Ofsted and the National Curriculum in England

The National Curriculum in England provides a framework for geography but doesn't specify the details of what should be taught or in what depth. Schools have the flexibility to choose their own curriculum approaches provided they pay sufficient attention to (a) context (b) structure (c) sequencing and (d) implementation. There is a significant emphasis on factual knowledge. Ofsted argue that it is essential to identify what children need to remember and to use it transferably in different circumstances.

The curriculum

The curriculum refers to what is taught and should support children to build their knowledge over time. Ofsted distinguish between different forms of knowledge:

- *Substantive knowledge* refers to knowledge relating to the themes and topics specified in the National Curriculum. This could be seen as the 'vocabulary' of geography.
- *Disciplinary knowledge* involves applying a geographical 'lens' to a particular or area of study and can be regarded as the 'grammar' of geography.

If pupils haven't grasped substantive geographical knowledge, then they will be unable to think or speak geographically.

Lesson planning

Ofsted argues that pupils get better in geography by building on their prior knowledge and applying it in new more complex ways (Ofsted 2021). Activities should therefore be selected which help pupils to build their knowledge and consolidate what they have already learnt in time-efficient ways. This draws attention to the importance of sequencing in geography. It is important to think carefully not only about the building blocks of geography but also about what comes after as well as what comes before any particular topic. In this way the curriculum becomes the assessment model.

Inspection findings

Inspection findings indicate that practice is not always as good as it could be. Areas of weakness include mapwork, fieldwork, sequencing and the application of geographical concepts (Ofsted 2021, 2023). To some extent this is unsurprising given that limited time has been spent learning how to teach geography during primary training (Ofsted 2023). However, there is also clear evidence that pupils enjoy geography and are curious about the world around them. The fact many are passionate about the Earth and the need to care for it also attracted very favourable comment from inspectors (Freeland 2021).

These prompts may help you prepare for an inspection:

- Identify a teacher who is responsible for developing the geography curriculum.
- Decide how geography will fit into your whole school plan.
- Make an audit of current geography teaching to identify gaps and weaknesses.
- Discuss and develop a geography policy which includes statements on overall aims, topic planning, teaching methods, progression, assessment and recording.
- See that all members of staff are familiar with the geography curriculum.
- Organise in-service training to rectify any areas of weakness.
- Review and update geography teaching resources.
- Devise an action plan for geography which includes an annual review procedure.
- Discuss the policy with the school governors.
- Provide a regular opportunity for discussing geography teaching in staff meetings.

Support and guidance

Primary Geography Quality Mark

The Primary Geography Quality Mark set up by the UK Geographical Association is a self-assessment framework designed to help subject leaders. There are three categories of award. The 'bronze' level recognises that lively and enjoyable geography is happening in your school; the 'silver' level recognises excellence across the school; and the 'gold' level recognises that excellence that is shared and embedded in the community beyond the school. The framework is divided into four separate cells: (a) pupil progress and achievement (b) quality of teaching (c) behaviour and relationships (d) leadership and management. For further details see the Geographical Association website.

Achieving accreditation for geography in school is a useful way of badging achievements and identifying targets for future improvement. This makes it an effective and efficient way of raising standards. The Geographical Association provides a wide range of support to teachers to help with this process. In addition to conferences and CPD sessions, it produces a journal for primary schools, *Primary Geography*, three times a year. See the Geographical Association website for full details of the books and guides it publishes for classroom use.

Networking, training and sharing ideas

Networking and sharing ideas can happen on an informal basis amongst friends and professional colleagues. Conferences and CPD training and events provide a more formal way of developing and extending your knowledge of geography teaching. The support that comes from networking will help you to grow in confidence and broaden your ideas. Speaking or writing about what you have been doing will further consolidate your ideas. Subject associations, environmental organisations and development education centres usually welcome new members with enthusiasm (for example, see the Geographical Association and the Royal Geographical Society websites). The sense of community that they foster cannot be underestimated.

References and reading

Barlow, Anthony, and Sarah Whitehouse (2019) *Mastering Primary Geography*, London: Bloomsbury Academic.

Bonnett, Alistair (2023) *What is Geography?* (2nd edn), Lanham, Maryland: Rowman and Littlefield.

Cannell, Jon (2023) 'Geographical concepts in primary education', *Primary Geography*, 112: pp8–9.

Catling, S. et al. (2022) 'Aspiring to High-Quality Primary Geography: A report on a study of the GA's Primary Geography Quality Mark Moderators' feedback to schools', Sheffield: Geographical Association.

Dolan, Anne M. (2020) *Powerful Primary Geography: A Toolkit for 21st-Century Learning*, London: Routledge.

Freeland, Iain (2021) 'Geography in outstanding primary schools', Ofsted: schools and further education & skills (FES) blog, gov.uk.

Ofsted (2021) 'Research review series: Geography', gov.uk.

Ofsted (2023) 'Getting our bearings: geography subject report', gov.uk.

Owens, Paula, Emily Rotchell, Sarah Sprake and Sharon Witt (2022) 'Geography in the Early Years: Guidance for doing wonderful and effective geography with young pupils', *Primary Geography*, 109: pp19–22.

Roberts, Margaret (2023) *Geography Through Enquiry: Approaches to teaching and learning in the secondary school* (2nd edn), Sheffield: Geographical Association (Chapter 4).

Scoffham, Stephen (ed.) (2016) *Teaching Geography Creatively (Learning to Teach in the Primary School Series)* (2nd edn), London: Routledge.

Scoffham, Stephen and Paula Owens (2024) *Bloomsbury Curriculum Basics: Teaching Primary Geography* (2nd edn), London: Bloomsbury.

Sprake, Sarah (2023) 'Geography's big ideas in the Early Years', *Primary Geography*, 112: pp10–12.

Tanner, Julia and Stephen Pickering (eds) (2017) 'Taking the Learning Outdoors at KS1', *Teaching Outdoors Creatively*, London: Routledge.

Trait, Georgie, et al. (2024) 'The key ingredients for quality geography', *Primary Geography*, 114, p19.

Willy, Tessa (ed.) (2019) *Leading Primary Geography: The essential handbook for all teachers*, Sheffield: Geographical Association.

Unit-by-unit notes

Unit 1: Earth in space

Lesson 1: EARTH, SUN AND MOON

The sun and the moon move through the sky above the Earth.
This lesson introduces children to Earth as an object in space. Like ancient people, young children often believe the world revolves around them. Thinking about the sun and the moon is one way of enlarging their understanding. It also raises questions about how we calculate and account for time. The sun is the measure of the length of the day. The phases of the moon divide the year into months. These are ideas are explored in subsequent lessons.

Using the story
The story aims to engage children emotionally. Like many young children, Baby Oscar is afraid of the dark. As you discuss the story, you might want to talk about how the moon orbits the Earth. This will help when children come to draw a picture or make a model.

Talking
Use the images on pages 4 and 5 for support. The sun is very hot, the Earth is covered with water, land and cloud, and the moon is lifeless and cold.

Amazing fact!
There is red hot rock beneath the Earth's surface.
The idea that there are red hot rocks beneath our feet challenges common-sense notions, so will be a surprise to some children.

For you to do
Check on the lunar calendar so you can prepare children for this activity. The new moon at dusk is a wonderful sight; so too is the full moon as it rises over the horizon.

Further activity
Ask children to make their own drawing of the Earth, sun and moon for a class display. They can then add labels.

Related picture books and stories
The Way Back Home, Oliver Jeffers, HarperCollins (2007)
Five Little Fiends, Sarah Dyer, Bloomsbury (2001)

> **Teacher's Guide photocopiable resource**
> Use page 30 to consolidate key concepts.
>
> **Workbook**
> See pages 2–3 for additional supportive activities.

Lesson 2: THE PLANETS

The Earth is one of eight planets which go round the sun.
The idea that the Earth orbits the sun was proposed by the Ancient Greeks, but was not widely accepted until the 17th century. Similarly, while the inner planets (Mercury and Venus) were known by the Babylonians, Neptune and other more distant parts of the solar system have only been discovered in the last few hundred years.

Using the story
This lesson introduces children to the solar system. The story plays on the difficulty children can have in counting – something which many of them will have encountered. The main teaching point is that Earth is one of eight planets, each with their own characteristics.

Talking
This activity introduces the idea of sequencing and distance. Use the diagram on pages 6 and 7 for support.

Amazing fact!
Jupiter has 95 moons.
You could refer back to the previous lesson and talk about our own moon and its qualities. Check that children understand that a moon orbits around a planet. Can children imagine what it would be like if the Earth had two or three moons rather than one?

For you to do
Children could work in pairs or groups to see which one remembers the most planets. Can they put the planets in order?

Further activity
Make a mobile of the planets for your classroom using discs of card, string and coat hangers for the frame.

Related picture books and stories
Adam's Amazing Space Adventure, Benji Bennett, Adam Printing Press (2009)
Where to Hide a Star, Oliver Jeffers, HarperCollins (2024)

> **Teacher's Guide photocopiable resource**
> Use page 31 to consolidate key concepts.
>
> **Workbook**
> See pages 4–5 for additional supportive activities.

Unit-by-unit notes

Lesson 3: DAY AND NIGHT

As the Earth spins in space, we get day and night.
The Earth spins on an axis marked by the North and South Poles. The equator is an imaginary line around the Earth halfway between the poles. Over the course of a year, all places on the Earth's surface experience the same number of hours of day and night, even though there are big seasonal differences. If you have access to a globe, it will reinforce the material and illustrations in the Pupil Book.

Using the story
Children may think that all creatures go to sleep at night, just like humans. The story makes the point that bats are able to find their way in the dark. Unlike humans, they use sounds to navigate. You might talk about other creatures that come out at night. You can also consider the contrasts between different times of day such as dawn and dusk.

Talking
Invite children to share their experience of day and night and how it structures our lives and what we do.

Amazing fact!
It never gets dark in the summer at the North Pole.
Introduce the idea that this is because the North Pole is tilted towards the sun. In winter the North Pole is tilted away from the sun, bringing continual darkness. *Going round the sun* (Pupil Book pages 28 to 29) explores how the tilt of the Earth causes the seasons.

For you to do
Discuss with children how it is important for all creatures, including humans, to experience darkness. Light can be a problem for some creatures, like the bat in the story. Light can also stop humans from sleeping well. Talk about how we can help (and save energy) by turning off lights we don't need.

Further activity
Make spinning tops using match sticks and circles of card. This will illustrate how the axis remains stationary while other parts of the card are in motion.

Related picture books and stories
The Light in the Night, Marie Voigt, Simon and Schuster (2019)
Apollo and the sun chariot (Greek myth)

Teacher's Guide photocopiable resource
Use page 32 to consolidate key concepts.

Workbook
See pages 6–7 for additional supportive activities.

Lesson 4: LAND AND WATER

Land and water cover the Earth's surface.
The astronauts who travelled to the moon in the 1960s and 1970s were the first people ever to view the Earth from space. The photographs they sent back captured both the fragility and unity of our world. Unlike other planets in the solar system, Earth is largely covered by water. Recognising the division between land and water provides the focus for this lesson.

Using the story
Water is all around us. Even when we dig into dry ground we will eventually come to saturated soil and rock, as the story illustrates. Talk with children about the places where they can find water locally. In what different ways have they already used water today?

Talking
This activity directs children's attention to the satellite image of the Earth on page 11. You could extend this by showing children a globe and asking them to look at how the continents are separated by oceans. How many oceans can they name?

Amazing fact!
Nearly all the water in the world is salty.
Discuss how seawater is salty because rivers carry minerals and salts into the ocean. These have built up over millions of years.

For you to do
Talk about the importance of throwing rubbish away carefully. Do children know about recycling? Discuss how rubbish can blow into drains and rivers, which lead into the sea. This can be dangerous for ocean creatures.

Further activity
Look carefully at a globe. Identify large blocks of land (continents) and islands around the coast. Get children to name some of them.

Related picture books and stories
Solving the Puzzle Under the Sea: Marie Tharp maps the ocean floor, Robert Burleigh, Simon and Schuster (2016)
The Crocodile Who Didn't Like Water, Gemma Merino, Macmillan Children's Books (2013)

Teacher's Guide photocopiable resource
Use page 33 to consolidate key concepts.

Workbook
See pages 8–9 for additional supportive activities.

Unit-by-unit notes

Unit 2: Planet Earth

Lesson 1: A LIVING PLANET

Water brings life to the Earth.
Although water covers large parts of the Earth's surface, most of it is salty. Fresh water, which is essential for humans and many other forms of life, represents only about 2.5% of the total water on the planet. Much of this lies deep underground or is frozen in mountain glaciers or the polar ice caps. As human numbers increase and demand increases, fresh water is becoming an increasingly scarce resource.

Using the story
The story of the little seed highlights something that many children will know: seeds need water to grow. The implication is that many plants will die if there are long periods without rain. However, too much rain may wash them away. It seems that climate change is bringing many places unpredictable weather of this kind. Talk with the children about any food plants they have grown.

Talking
As you talk about the photos, encourage children to think about other creatures which are also part of river, marsh and ocean habitats. Why is water important to them?

Amazing fact!
Some deserts have no rain for more than 100 years.
Talk about how deserts are found in places that have very dry winds. Sometimes moist air brings rain.

For you to do
The practical experience of growing a plant from a seed is a valuable and lasting lesson. Children could record how it changes.

Further activity
Plant some mustard and cress seeds on a damp cloth to see how quickly they germinate and how they respond to water, warmth and light.

Related picture books and stories
Eliza and the Moonchild, Emma Chichester Clark, Andersen Press (2007)
Once Upon a Raindrop: The story of water, James Carter, Caterpillar Books (2018)

> **Teacher's Guide photocopiable resource**
> Use page 34 to consolidate key concepts.
>
> **Workbook**
> See pages 10–11 for additional supportive activities.

Lesson 2: THE SHAPE OF THE LAND

The land consists of mountains, hills and lowlands.
The Earth's surface is constantly changing. As the tectonic plates move against each other, mountains are built up and volcanoes come to the surface. At the same time water, wind, ice and other forces wear away the land and try to reduce it to sea level. The interaction of these processes has created the landscapes we see today. Ensure that children understand that water flows downhill, so rivers flow from the mountains into the sea.

Using the story
The story introduces some of the features found along the course of a river. These are highlighted in the 'Read the words' list. Getting children to write these words alongside simple drawings in their geography notebooks is a good way to build their understanding.

Talking
Use the 'Read the words' box to support this activity. You might need to explains some of the words. 'Current' is often misunderstood and confused with the homophone 'currant'!

Amazing fact!
The world's longest river is called the Nile. It is over 6000 kilometres long.
See if children can find the Nile and Africa on a world map. What other big rivers can they find?

For you to do
Ensure children understand that things that go down the drain ends up in rivers. Discuss how toothpaste can be dangerous for the creatures that live in our rivers, so we should only use what we need.

Further activity
Find out about the longest river and highest mountain in (a) your country (b) the world.

Related picture books and stories
River Story, Meredith Hooper, Walker Books (2015)
Hike, Pete Oswald, Walker Books (2021)

> **Teacher's Guide photocopiable resource**
> Use page 35 to consolidate key concepts.
>
> **Workbook**
> See pages 12–13 for additional supportive activities.

Unit-by-unit notes

Lesson 3: VOLCANOES

Volcanoes bring hot rocks to the surface from deep underground.
There are approximately 1500 active volcanoes in the world. Around three-quarters of these are in what is known as the 'ring of fire' around the shores of the Pacific Ocean. Here, Earth movements bring hot rocks from deep underground up to the surface. Children of all ages are fascinated by volcanoes. There will be opportunities to support this lesson with up-to-date images and videos from the internet. You may also find reports of eruptions which are happening at the current time.

Using the story
This tale is a gentle way of introducing a dramatic topic. The plumes of smoke, poisonous gas and red-hot lava that accompany a volcanic eruption will give children plenty to talk about. You may choose to link this to creative writing and/or talking, as volcanoes are dramatic landscapes which many children find interesting.

Talking
Take the opportunity to introduce some specific words and terms such as crater, lava, cone and vent as you talk with children about volcanic eruptions.

Amazing fact!
Over millions of years, volcanoes have helped to create the air we breathe today.
Talk about how the volcanoes which erupted in the past let out so much steam and gas that it changed the air around us (the atmosphere).

For you to do
Young children can sometimes think that people, rather than natural forces, created mountains, hills and valleys. Talk with the children about how they think the landscape in your area, like the nearest hill, was formed.

Further activity
Show the children some examples of volcano paintings and ask them to make one of their own.

Related picture books and stories
Gopher to the Rescue! A volcano recovery story, Terry Catasus Jennings, Abordale (2012)
Vladimir the Volcano: A tale of an unforeseen eruption, Rana Boulos, Nature Speaks series (2021)
Vulcan and his underground forge (Roman myth)

> **Teacher's Guide photocopiable resource**
> Use page 36 to consolidate key concepts.
>
> **Workbook**
> See pages 14–15 for additional supportive activities.

Lesson 4: WORLD WONDERS

There are many special sights to see in the world.
The seven wonders of the ancient world have found their place in history. Today people seek to protect landscapes and places which they particularly value by establishing nature reserves, national parks and heritage sites. Recognising that we live on a remarkable and very special planet is the central idea behind this lesson. Getting children to celebrate different 'wonders' is an essential basis for caring about them in later life.

Using the story
This story features 'wonders' above ground (icebergs) and below ground (caves), as well as temporary phenomena (the northern lights). Children learn about elements such air, rock and water from Book 1 *World around me*. This lesson builds on that earlier learning.

Talking
You could guide children to focus on ancient wonders or modern buildings as well as landscapes, plants and creatures. You could look at pictures of these wonders to support the discussion.

Amazing fact!
Some icebergs contain water that fell to Earth thousands of years ago.
Talk about how the ice sheets in polar regions are formed from snow which fell in the past. This tells scientists about the climate at that time.

For you to do
The things which children think are special or unusual about their local environment are often very different from the things which adults select. Ask them to give reasons for their choices.

Further activity
Collect some photographs of different 'wonders' for a class display.

Related picture books and stories
Wonder Walkers, Micha Archer, Nancy Poulson Books (2021)
Dear Earth, Isabel Otter, Caterpillar Books (2021)

> **Teacher's Guide photocopiable resource**
> Use page 37 to consolidate key concepts.
>
> **Workbook**
> See pages 16–17 for additional supportive activities.

Unit-by-unit notes

Unit 3: Weather and seasons

Lesson 1: EXPERIENCING THE WEATHER

There are many different types of weather.
The sun provides the energy which drives the world's weather. The warmth of the air at the equator contrasts with low temperatures at the poles. The wind seeks to equalise these differences in temperature and pressure. However, as the wind blows from place to place, it also carries moisture which can result in cloud and rain. We experience these changing conditions in daily and seasonal weather patterns.

Using the story
The story illustrates how the weather can be changeable. Mika complains that the weather 'isn't fair', but can the weather ever be fair? Discuss this. Do they know of any places that always seem to have 'good' or 'bad' weather?

Talking
Talk about how the weather affects our lives. Can children think of any times when they have been in a difficult situation because of the weather? How does the weather affect the jobs that some adults do?

Amazing fact!
It almost never rains in Antarctica.
Discuss how Antarctica is so cold that it almost always snows when the weather is wet (and snow is frozen water).

For you to do
Support this activity by discussing why some places always seem windy, whereas others are not windy. Talk about why this is.

Further activity
Set up a display table with clothes and items for different types of weather.

Related picture books and stories
Rainy Day, Emma Haughton, Doubleday (2000)
The North Wind and the Sun, Brian Wildsmith (2007)

> **Teacher's Guide photocopiable resource**
> Use page 38 to consolidate key concepts.
>
> **Workbook**
> See pages 18–19 for additional supportive activities.

Lesson 2: DIFFERENT TYPES OF WEATHER

We can describe the weather using words and symbols.
The UK, along with other temperate climate regions, experiences a great range of weather types. Although extreme weather is unusual in the UK, the weather often changes within just a short timescale. As a result, people are often interested in what is going to happen next and pay attention to weather forecasts. Forecasters have devised a range of weather symbols to communicate quickly and effectively. Some of these symbols are introduced in this lesson.

Using the story
This story builds on the previous lesson (Pupil Book pages 20 to 21) by introducing a range of different weather types. Can children think of any other examples, such as frost, hail, drizzle, etc.?

Talking
Prompt children by thinking of all the different words they know to describe hot and cold weather. Can they arrange these words in a sequence?

Amazing fact!
The water in a rain drop never ever disappears.
Discuss how all the rainwater eventually makes its way into rivers and flows back into the oceans.

For you to do
Lots of people nowadays, especially in urban areas, tend to spend a lot of time in heated or air-conditioned environments. Talk about how we can change our clothes to match the weather. Not all children will naturally think about this.

Further activity
Make pictures and weather symbols to show the different types of weather on Mika's holiday.

Related picture books and stories
Meet the Weather, Caryl Hart, Bloomsbury (2023)
Shinoy and the Chaos Crew: Does rain ever fall up?, Claire Llewellyn, Collins (2021)

> **Teacher's Guide photocopiable resource**
> Use page 39 to consolidate key concepts.
>
> **Workbook**
> See pages 20–21 for additional supportive activities.

Unit-by-unit notes

Lesson 3: EXTREME WEATHER

Sometimes the weather can be wild and exciting.
We are becoming increasingly aware of extreme weather incidents. Better communications and increases in population mean that disasters are both more likely to happen and better publicised when they do. At the same time, climate change is resulting in more unpredictable weather patterns and is making extreme weather more severe. This is a trend which is likely to continue. This lesson considers the power of nature but tries to avoid being alarmist for young children. Extreme weather can be dramatic without being frightening.

Using the story
The story is structured around the well-known saying 'Be careful what you wish for'. It makes a serious point in a light-hearted way. Extreme weather disrupts people's lives. As well as filling homes with water, it can wash away bridges, disrupt electricity and food supplies and, ironically, lead to shortages of drinking water.

Talking
You could discuss how flood water would damage your classroom. It would cover the floor, soak the walls, damage decorations and make the electricity unsafe.

Amazing fact!
There are millions of thunderstorms around the world each year.
Talk about how lightning occurs when electricity flows between storm clouds or from clouds to the ground. Thunder is the sound this creates.

For you to do
As they engage with this activity, encourage children to think about what happened before and after each photo was taken.

Further activity
Consider any real-life extreme weather events that may be happening around the world as you are teaching this lesson. Discuss these with the children in an appropriate way, so as to inform but not scare them.

Related picture books and stories
After the Storm, Nick Butterworth, HarperCollins (2011)
Wild Weather, Chris Oxlade, Collins (2015)

Teacher's Guide photocopiable resource
Use page 40 to consolidate key concepts.

Workbook
See pages 22–23 for additional supportive activities.

Lesson 4: SEASONS

There is a pattern of seasons during the year.
The UK, along with many other countries in mid-latitude locations, has four seasons of roughly equal length. The changes in weather and variations in the length of the days and night from season to season has become part of everyday life and culture. Seasonal celebrations mark out the year. The school calendar still follows the pattern of the seasons, as do many sporting events.

Using the story
The story highlights the rhythm of the seasons. It also demonstrates the importance of thinking about the future. Plants and animals prepare for harsh winter weather when they have the opportunity earlier in the year. Our bodies also prepare and respond to seasonal changes.

Talking
Talk about the pattern of the school year and how it is linked to the seasons. How are different seasons celebrated in your school?

Amazing fact!
In the UK there can sometimes be winter weather even in the middle of summer.
Ensure children understand that the seasons are a pattern of weather that happen over a number of months. The weather changes from day to day.

For you to do
Use the photos to prompt children to think back over the year and their personal experiences.

Further activity
Make some seasonal drawings for a class display or collage.

Related picture books and stories
Persephone and pomegranate seeds (Greek myth)
The Selfish Giant, Oscar Wilde (1888)

Teacher's Guide photocopiable resource
Use page 41 to consolidate key concepts.

Workbook
See pages 24–25 for additional supportive activities.

Unit-by-unit notes

Lesson 5: GOING ROUND THE SUN

The seasons change as the Earth goes round the sun.
The previous lesson established the pattern of the seasons. This lesson explores what causes them. There is one key point: the Earth's axis, rather than being vertical, is tilted. This means that over a period of a year, places are tilted first towards and then away from the sun. This is a complex idea which is best illustrated through models and diagrams. The fact that the sun is higher and lower in the sky at different times of the year will be appreciated by children. *For simplicity, the diagram refers to seasons in the northern hemisphere. The southern hemisphere has the opposite seasons.*

Using the story
The story introduces some of the misconceptions children may have about what causes the seasons. This will reveal the extent of their prior knowledge before they look at the diagram. If possible, reinforce understanding by showing how the Earth's axis is tilted using a globe.

Talking
Explain to children that the North Pole always has daylight in the summer and darkness in the winter. Can they find the position of the North Pole in the diagram to see why this happens?

Amazing fact!
Mercury takes just 88 days to go round the sun. Neptune takes 165 years.
Do children remember which planet is closest to the sun? Refer back to *The planets* (Pupil Book pages 6–7). Discuss how Mercury is the closest to the sun so takes the shortest time to go round it. Neptune is furthest away, and so takes the longest.

For you to do
Discuss how some birds and creatures struggle to find food in the winter. You can explain that some put on fat in the autumn as a precaution, and others hibernate.

Further activity
Spin a globe slowly, focusing on the North Pole. Discuss how it has 24 hours of sunshine in summer and 24 hour darkness in winter.

Related picture books and stories
Earth Song, Susan Reed, Barefoot Singalongs (2024)

Teacher's Guide photocopiable resource
Use page 42 to consolidate key concepts.

Workbook
See pages 26–27 for additional supportive activities.

Unit 4: Local areas

Lesson 1: SHELTER

Homes give us warmth and shelter.
All creatures need protection from the weather. They also need to keep safe from predators. Humans have responded to these basic needs in different ways. Cave dwellings were a ready-made solution. Castles provided a stronghold where many people could defend themselves at times of unrest. For most people, four walls and a roof represent security and a place where they feel they belong. This lesson focuses on tents to highlight the importance and features of a shelter. Settlement – a key geographical idea – is then developed in the lessons on streets and villages.

Using the story
The story highlights how shelters, including tents, provide protection from the elements. Wind, rain and cold are perhaps the main threats. Discuss these.

Talking
Children may have been camping or read about camping in stories. What do children think they would like and dislike about camping? Discussing where to pitch a tent raises further questions about site and facilities.

Amazing fact!
People have been living in tents from the earliest times in history.
Discuss how some people live in tents today. Tents provide simple homes for herders as they move from place to place with their animals. Others use tents for a short time until they have somewhere permanent to live.

For you to do
Draw attention to the differences between a house and a home. A house is simply a structure. A home is where people live and what they do to make it homely.

Further activity
The children could make a shelter for a small animal such as a mouse or a lizard using modelling clay.

Related picture books and stories
In Every House on Every Street, Jess Hitchman, Little Tiger Press (2020)

Teacher's Guide photocopiable resource
Use page 43 to consolidate key concepts.

Workbook
See pages 28–29 for additional supportive activities.

Unit-by-unit notes

Lesson 2: HOUSES AROUND THE WORLD

People build houses in lots of different ways.
Traditionally houses were built from local resources such as wood, stone, thatch and clay (bricks). The Industrial Revolution and the development of roads and railways have made it possible to transport heavy building materials from one place to another. Newer materials such as plastic, steel, glass and concrete have heralded further developments in domestic architecture. Air conditioning and central heating help to control the temperature inside. Water, energy and communication systems provide links to the external surroundings. These all need resources to keep them running, which adds to the 'environmental footprint'. One of the challenges today is to see how we can live in comfort without putting undue strain on our surroundings.

Using the story
The story compares living in the country with city life. This opens up questions about the advantages and disadvantages of different types of settlement. Comparing buildings is one way to show differences. For example, blocks of flats/apartments are an urban phenomenon.

Talking
A discussion around building materials provides a natural link to science. You can also link to history because the houses in the drawings come from different periods in the past. Relate the lesson to children's own experiences. What different house types do they know?

Amazing fact!
The world's tallest buildings are as high as many mountains.
Some famous examples of tall buildings are the Burj Kalifa and the Shard. Talk about how tall buildings are significant features which help us find our way. What landmarks are there in your area?

For you to do
Young children often find it hard to understand why there are different lines in an address. Explain that the first line relates to the house and street, the second line to the village or town and the third to the region. Postcodes encapsulate all this information in a simple combination of numbers and letters.

Further activity
Go for a short walk to look at houses in your area. How many different types of building material can you find?

Related picture books and stories
The House at Pooh Corner, A. A. Milne, Egmont (1928)
Dani Binns: Brilliant Builder, Lisa Rajan, Collins (2020)

> **Teacher's Guide photocopiable resource**
> Use page 44 to consolidate key concepts.
>
> **Workbook**
> See pages 30–31 for additional supportive activities.

Lesson 3: LIVING IN A VILLAGE

People live together in groups or communities.
A village is a basic form of settlement which from very early times has enabled people to settle in one place instead of roaming. Villages need a water supply for drinking and washing and fields for growing crops and grazing animals. They also provide the social benefits of a community. In the past, people living in villages specialised in tasks and provided essential services (for example blacksmiths and bakers), so villages were often self-sufficient. With industrialisation, they have often now become desirable places to live.

Using the story
Ewan visits some key places in a typical UK village, such as the shop and the school. Talk about whether a village needs to have these sorts of places to be a community.

Talking
Just as a house becomes a home when people live in it, a village is a living community. As well as play areas, there is sometimes a community park. Talk about how these places bring people together.

Amazing fact!
Nearly every village in the United Kingdom is hundreds of years old. A village in the UK always has a church.
Talk about why children think some villages have gone on growing whilst others have stayed the same size.

For you to do
It is important for children to know their address in case they get lost. This activity makes them think about how other people know where they live.

Further activity
Make a class display of a village. This could either consist of pictures of buildings and features arranged around a street plan or models set out on a display table.

Related picture books and stories
Katie Morag Delivers the Mail, Mairi Hedderwick, Red Fox (2010)

> **Teacher's Guide photocopiable resource**
> Use page 45 to consolidate key concepts.
>
> **Workbook**
> See pages 32–33 for additional supportive activities.

Unit-by-unit notes

Lesson 4: EXPLORING LOCAL STREETS

There are lots of things to see in a street for people to use.
The items which are added to a street to make it habitable are known as street furniture. There is a great variety of street furniture including benches, bus stops, post boxes, road signs and litter bins. This lesson draws attention to the range of items in a street and helps children to understand why each one is needed. This lesson could be supported by local fieldwork in a street near you, if possible.

Using the story
Many children will have bumped into something as they walk down the street. Talk about what other things might have got in Isla's way on the street. Are there any items of street furniture which children think are unnecessary?

Talking
The photograph of the drain (photograph D) is a reminder that streets are used to connect houses to a range of different services. Talk about how the things we put down the drain do not simply disappear. Traps and covers are clues to these underground systems.

Amazing fact!
The first street lights were in China 2500 years ago.
Ask children what might happen to them if the wires carrying electricity stopped working.

For you to do
Discuss what would happen if each of the items of street furniture was not there. Why are they useful?

Further activity
Collect photographs for a display about the things in the streets around your school. See if you can find something for different letters of the alphabet.

Related picture books and stories
Belonging, Jeannie Baker, Walker Books (2004)
The Green Line, Polly Farquharson, Frances Lincoln (2009)

> **Teacher's Guide photocopiable resource**
> Use page 46 to consolidate key concepts.
>
> **Workbook**
> See pages 34–35 for additional supportive activities.

Lesson 5: UNDER YOUR FEET

There are lots of pipes and wires under the pavement.
Streets are full of clues to the systems that sustain modern lifestyles. Water, electricity and gas are supplied to most homes, and rubbish and waste are removed promptly and efficiently. These services link us to places beyond the immediate environment which we depend on for our survival. Without them it would soon become impossible for people to live clustered together in towns and cities.

Using the story
In a humorous way, the story makes the point that essential services are supplied to our houses through underground pipes and wires. The traps and gratings in the pavement provide evidence of different services which children can see on a local environmental walk.

Talking
Public utilities need to be constantly maintained and upgraded. Discuss the work that children have seen being done to keep them running smoothly. Is it always necessary to bury pipes and wires under the ground?

Amazing fact!
The amount of water we use in a week weighs about the same as a small car.
Talk about why water is so important in our lives. What do we use it for?

For you to do
Talk about how turning off lights when we do not need them is one way children can reduce their 'environmental footprint'. What else could they do?

Further activity
Make a set of rubbings of lids and covers in the pavements around your school.

Related picture books and stories
Superworm, Julia Donaldson, Alison Green Books (2012)
The Street Beneath My Feet, Charlotte Guillain, Words and Pictures (2017)

> **Teacher's Guide photocopiable resource**
> Use page 47 to consolidate key concepts.
>
> **Workbook**
> See pages 36–37 for additional supportive activities.

Unit-by-unit notes

Unit 5: Maps and plans

Lesson 1: MAPS AND STORIES

Picture maps can show us about the places featured in songs and stories.
Nearly every story is set in a place or location and many involve a journey of some kind or other which can be mapped. Maps take a wide variety of forms. They are best described to young children as pictures of places. This lesson introduces children to maps through stories, which relate to their interest and understanding. You might even decide not to use the word 'map' but to talk about 'pictures' instead.

Using the story
This story is a retelling of Aesop's fable of the Hare and the Tortoise. You could get children to act it out before they draw their own picture maps of the race. What were the key stopping points along the way? Do they want to add any additional features of their own?

Talking
Make a list of the stories the children suggest. This will be useful if you want to develop the lesson by making picture maps to go with stories they know and like.

Amazing fact!
People started making maps before they learnt to read and write.
Ask children how they know how to get from one place to another. Talk about why maps are useful.

For you to do
Ask children to show three places mentioned by the person they interviewed on their story map.

Further activity
Make a class display to illustrate the places and events in a different story or song.

Related picture books and stories
The Hare and the Tortoise, Brian Wildsmith, Oxford (2007)
Rosie's Walk, Pat Hutchins, Bodley Head (2018)

> **Teacher's Guide photocopiable resource**
> Use page 48 to consolidate key concepts.
>
> **Workbook**
> See pages 38–39 for additional supportive activities.

Lesson 2: TREASURE ISLAND

We use maps to show places both real and imagined.
Maps have great power because they are a record of the lie of the land and its key features. Identifying these helps us to travel and make journeys without getting lost. In this lesson children explore how maps and compass directions enable us to locate places and objects – a key geographical skill. The idea of finding buried treasure will stimulate imaginative responses. As you talk about the map, you might think about treasures in a broader sense to include buildings, plants and landscapes.

Using the story
Toya and Efe use compass directions to help them explore the treasure island. Rehearse the four points of the compass. Give children the chance to use a compass for themselves if possible. Can they turn to the west and look to the south? What is in the north of the classroom?

Talking
The activity introduces children to alphanumeric grids. Explain that in this system, the letter comes first and the number comes second. A simple way to remember this is that you go 'along the corridor and up the stairs'.

Amazing fact!
There are around 25 000 islands in the Pacific Ocean.
Discuss how islands can be different sizes, small and large. Build on understanding by explaining that islands can be any piece of land, smaller than a continent, that is entirely surrounded by water. Do children know any islands?

For you to do
Making a map of the journey to school is a classic geography exercise. Tell children in advance to note the things they pass on their daily journey. This will be easier for those who cycle or travel on foot. As an alternative they could map a journey round the school grounds.

Further activity
Get children to make their own map of an imaginary island or place. Some might be able to devise directions for finding treasure.

Related picture books and stories
The Treasure of Pirate Frank, Mal Peet and Elspeth Graham, Nosy Crow (2017)

> **Teacher's Guide photocopiable resource**
> Use page 49 to consolidate key concepts.
>
> **Workbook**
> See pages 40–41 for additional supportive activities.

Unit-by-unit notes

Lesson 3: DIFFERENT PLANS

Plans show the shape of things and places.
While maps show large areas in a generalised way, plans show small areas in great detail. Plans are particularly important to builders and architects, who use them to visualise what they are constructing. Children can be introduced to plans from an early age, particularly if the plans show places or things they already know. Contrasting the side view (elevation) with the overhead view (plan) can be an interesting extension activity.

Using the story
Discuss why the children in the story were unhappy with the new school, apart from those in Class 1. Now look at the pictures and plans of the pond, school and other places shown in the book. Can children work out how each plan relates to the place it portrays?

Talking
You could introduce this activity by getting children to draw a plan of a familiar place such as the playground. A flowerbed or small garden is another option.

Amazing fact!
The buildings on Palm Island in Dubai are arranged in the shape of a palm tree.
The plan (or bird's eye view) uses an unfamiliar perspective. Show children lots of examples of plans to develop their understanding.

For you to do
If children find it hard to make a plan of their classroom, get them to make a plan of items arranged on their desk or a display table first. This will give them the chance to look down on what they are drawing.

Further activity
Select some familiar items such as bottles, cartons and containers. Ask children to draw round the base to create a plan. See if they can they match the items to their plans once they have been separated.

Related picture books and stories
Percy's Bumpy Ride, Nick Butterworth, HarperCollins (2011)
Martha Maps It Out, Leigh Hodgkinson, Oxford (2022)

> **Teacher's Guide photocopiable resource**
> Use page 50 to consolidate key concepts.
>
> **Workbook**
> See pages 42–43 for additional supportive activities.

Lesson 4: THE VIEW FROM ABOVE

Plans show what places look like from above.
Although there are no set rules, formal maps and plans are often drawn from above. Taking an overhead view solves the problem of perspective, but the shapes revealed are often unfamiliar. With increasing height, the representation tends to become even more abstract. Making comparisons with overhead photographs helps to bring back a measure of 'reality'.

Using the story
What other places and things have a square overhead shape? Get children to set up play equipment to show a traffic jam, a farm or a street scene to help them answer this question.

Talking
The houses are rectangular in shape, the factory looks a bit like a 'Z', the sports ground has oval ends and the tree shadows look triangular.

Amazing fact!
Astronauts can see the pyramids in Egypt from space.
Show children satellite images of the Earth from space to see what features show up. The NASA website has lots of examples.

For you to do
If there is a park near your school, children could visualise the shapes they would see from above. Look at a large-scale map to gain an idea of the street pattern and other overhead shapes.

Further activity
Compare maps and aerial photographs of your school and its surrounding area.

Related picture books and stories
A Balloon for Grandad, Nigel Gray, Orchard Books (2002)
Zoom, Istvan Banyai, Viking Kestrel (2007)

> **Teacher's Guide photocopiable resource**
> Use page 51 to consolidate key concepts.
>
> **Workbook**
> See pages 44–45 for additional supportive activities.

Unit-by-unit notes

Unit 6: The United Kingdom

Lesson 1: COUNTRIES AND CAPITALS IN THE UNITED KINGDOM

There are four countries in the United Kingdom. Each country has a capital city.
The United Kingdom consists of four countries which have amalgamated over centuries. England and Wales were brought together in 1284, Scotland joined the union in 1707, and Northern Ireland was added in 1921. Thinking of the United Kingdom as separate countries sometimes creates confusion. This lesson aims to help children identify different areas, and name and locate their respective capital cities.

Using the story
Discuss where Arjun must have started his imaginary journey if he lives in the United Kingdom. What is special about each of the places he imagines? Explain that the Giant's Causeway is a famous place with around 40 000 interlocking rock columns. Can children think of some other things Arjun might have liked on his imaginary tour?

Talking
Some children will know about the sports teams which represent the different UK countries. Others may have visited different regions for their holidays. Or they can simply look at the map on page 49 to compare the size of the four countries. Which country has the most islands? Can they name their capital cities?

Amazing fact!
Great Britain (England, Wales and Scotland) is the ninth largest island in the world.
Look at some maps and make a collection of island shapes. Can they find any coastlines which are completely straight? (*The coastline is never straight!*).

For you to do
You could show children photos of the different countries and their capitals as they undertake this activity.

Further activity
Divide the children into four groups – one for each country of the UK. Get them to find out all the things they can about their area.

Related picture books and stories
We Are Britain!, Poems by Benjamin Zephaniah, Frances Lincoln (2003)
Coming to England, Floella Benjamin, Macmillan (2020)

> **Teacher's Guide photocopiable resource**
> Use page 52 to consolidate key concepts.
>
> **Workbook**
> See pages 46–47 for additional supportive activities.

Lesson 2: MOUNTAINS, RIVERS AND SEAS IN THE UNITED KINGDOM

The United Kingdom has a remarkably varied landscape.
The United Kingdom has a varied landscape. Scotland is famous for its mountains and lochs. England has lowlands and hills. Wales is known for its valleys and Northern Ireland for its green fields. The rocks beneath the surface are one of the reasons for these different landscapes. The erosion that occurred during the last ice age is another factor.

Using the story
The story focuses on three different landscapes: mountains, rivers and lowlands. Talk about what makes them distinctive. Have any children been climbing in the mountains of Wales or Scotland? Have any of them visited the Norfolk Broads or the River Severn? What other places do they know and what makes them special?

Talking
You could extend this activity by getting children to compare the landscape features of the different UK countries. They could also practise using compass directions. For example, the Mourne Mountains lie to the west of the Lake District across the Irish Sea.

Amazing fact!
Some mountains in the United Kingdom were once active volcanoes.
Talk about how the landscape has changed a lot over time. What we see today provides clues to what our planet was like in the past.

For you to do
This activity directs children to think about the landscape where they live. Even if they live in a big town or city, there are likely to be hills and rivers in places.

Further activity
Create a wall display about the UK. Get children to add drawings of different features to a poster or outline map.

Related picture books and stories
ABC UK, James Dunn, Frances Lincoln Children's Books (2009)
Mighty Mountains, Swirling Seas, Valerie Bloom, Collins (2015)

> **Teacher's Guide photocopiable resource**
> Use page 53 to consolidate key concepts.
>
> **Workbook**
> See pages 48–49 for additional supportive activities.

Unit-by-unit notes

Unit 7: Different environments

Lesson 1: LIVING IN THE ARCTIC

The Arctic is the region around the North Pole. It is very cold and snowy.
The Arctic is one of the coldest and harshest regions on Earth. Around the North Pole the ocean is permanently covered in ice. Although the ice is getting thinner, it still extends thousands of kilometres in the winter months. Around the shores of the Arctic Ocean there are marshes, lakes and peat bogs, which form a frozen wilderness known as the tundra. Creatures such as the polar bear have evolved to survive in these conditions. However, now that temperatures are rising because of global warming, the ice is melting and these creatures are losing their habitat. This decline can be halted. For example, international action to phase out the gases that were causing the ozone hole shows what is possible.

Using the story
The story introduces some of the iconic animals that live in the Arctic. What do children know about them? What other Arctic creatures can they think of?

Talking
Introduce how the Arctic is extremely cold, has extreme winds and extreme variations in daylight. Explorers have been attracted by the challenge of these conditions. Tourists like to go there to see the northern lights.

Amazing fact!
There is water, not land, under the ice at the North Pole.
Talk about how the Arctic sea ice is around two to three metres thick (demonstrate with a ruler). It grows in winter to cover an area that is larger than Europe.

For you to do
Modern heating and air conditioning help people with the weather conditions outside. This activity is a reminder of the environmental benefits of adjusting our clothing.

Further activity
Find out about other creatures that live in the Arctic. Gather photographs and drawings for a display.

Related picture books and stories
Polly and the North Star, Polly Horner, Orion Children's Books (2002)
Arctic Life, Sean Callery, Collins (2013)

> **Teacher's Guide photocopiable resource**
> Use page 54 to consolidate key concepts.
>
> **Workbook**
> See pages 50–51 for additional supportive activities.

Lesson 2: LIVING IN THE RAINFOREST

The rainforest lies on the equator. It is hot and wet.
Tropical rainforests are hot and wet with intensive tree cover. This environment provides a habitat for a multitude of life. Although the rainforest only covers 6% of the Earth's land surface, it contains more than half of the world's plant and animal species. The trees also put so much moisture into the air that they regulate the weather. Cutting down the rainforest is not only an ecological disaster, but scientists fear irreversible climate consequences. Working with local communities offers a promising way forward.

Using the story
The main theme of this story is the variety of plant and animal life in the rainforest. Ask children to list the creatures Isla and Ewan come across on their walk. It is of course an imaginary tale as such a range of creatures could never been seen on a short stroll – and it is not sensible to walk near dangerous animals like jaguars!

Talking
This activity is a reminder that knowing specific vocabulary can help children express ideas. In addition to plants and animal life, encourage children to think of words which describe the weather.

Amazing fact!
The rainforest is home to half of the world's plant and animal species.
Discuss why it is important to look after the rainforest. It will save not just trees but many plants and animals – including some which scientists haven't yet studied.

For you to do
It's extraordinary to think that acorns and other seeds can turn into tall trees. This is an interesting conversation for children to have with an adult.

Further activity
Fold a piece of paper in half like a greetings card. Children can draw rainforest trees and plants on the front, and rainforest creatures inside.

Related picture books and stories
We're Roaming in the Rainforest: An Amazon adventure, Laurie Krebs, Barefoot Books (2010)

> **Teacher's Guide photocopiable resource**
> Use page 55 to consolidate key concepts.
>
> **Workbook**
> See pages 52–53 for additional supportive activities.

Unit-by-unit notes

Lesson 3: LIVING IN THE DESERT

Most deserts are very hot and dry.
Australia is the driest inhabited continent. Deserts make up most of the outback in the central regions. In the past, deserts were seen as inhospitable places where few people can survive. Now tourists can visit the amazing scenery and see the rocks, gorges and coloured sands using powerful modern vehicles. Finding a balance between allowing people to explore these vulnerable habitats and protecting them from disturbance is an interesting challenge.

Using the story
Talk about what children think makes deserts special. As well as heat and lack of water, they might mention the brilliant sunlight and sandstorms. Extend this by reminding them that many deserts are rocky rather than sandy. Some, like the Mongolian desert, are also extremely cold.

Talking
You could extend this activity by getting children to draw a picture sequence of Tami's adventure.

Amazing fact!
Deserts cover one third of the land on Earth.
Talk about how some creatures like kangaroos are found naturally in (are native to) Australia. They have adapted to live in the dry, hot conditions.

For you to do
This activity draws attention to the importance of soil. Although it is at the bottom of the food chain, it is the cornerstone of life on Earth. Talk about how the plants provide food for creatures.

Further activity
Create a desert environment using a sand tray and small drawings and models of different plants and creatures.

Related picture books and stories
Just So Stories (How the Camel Got His Hump), Rudyard Kipling, Oxford (1902), (2009)
In the Bush: Our holiday at Wombat Flat, Roland Harvey, Allen and Unwin (2007)

Teacher's Guide photocopiable resource
Use page 56 to consolidate key concepts.

Workbook
See pages 54–55 for additional supportive activities.

Lesson 4: ANIMALS AROUND THE WORLD

We share our world with many different plants and creatures.
Over 4000 million years a great diversity of plant and animal life has evolved on Earth. By the time humans first emerged, biodiversity was probably at its peak. Since then, human activity and the consequent loss of habitats has gradually taken its toll. In recent years there has been a massive decline in wildlife, which has plummeted by over 70% in just 50 years. This alarming trend means we are now living in a period of mass extinction of geological significance. Helping children to celebrate the diversity of the world may encourage them to want to care for it as they grow older.

Talking
This will help to consolidate children's knowledge of the key world environments in the previous three lessons.

Amazing fact!
The blue whale is the largest creature on Earth. It is as heavy as 30 elephants.
Talk about how scientists know about over a million types of animal (species). But they haven't yet found out about many more animals that live on our planet.

For you to do
As children learn more about the threats to wildlife it is important to give them the chance to do something positive to protect it.

Related picture books and stories
If the World Were 100 Animals, Miranda Smith, Red Shed (2022)

Teacher's Guide photocopiable resource
Use page 57 to consolidate key concepts.

Workbook
See pages 56–57 for additional supportive activities.

Unit 8: World maps

Continents and oceans are very large units which provide a general framework that enables us to make sense of the world. However, it is difficult to be precise about where an ocean begins and ends. Continents too tend to merge into each other. Islands create special difficulties. Children can also find it difficult to distinguish between countries and continents.
You can choose to use the maps in Unit 8 as separate lessons to help children make sense of the basic divisions. You can also use these as a reference throughout the lessons.

Teacher's Guide photocopiable resource
Use pages 58–59 to consolidate key concepts.

Workbook
See pages 58–61 for additional supportive activities.

Photocopiable resource matrix

Photocopiable resource	Description
Earth in space	
1 Earth, sun and moon	Children colour pictures of the Earth, sun and moon and add labels.
2 The planets	Starting with a key, children colour the planets.
3 Day and night	Children colour a globe and draw pictures to show day and night.
4 Land and water	A simple exercise in which children identify land and water on a globe.
Planet Earth	
5 A living planet	Children colour and order drawings to show a sequence in plant growth.
6 The shape of the land	Children colour a landscape picture and turn it into a jigsaw puzzle.
7 Volcanoes	A practical activity in which children create their own model volcano.
8 World wonders	Children colour two drawings of world wonders before drawing a picture of their own.
Weather and seasons	
9 Experiencing the weather	Children label and colour drawings of different types of weather.
10 Different types of weather	Children colour weather symbols and cut them out for a game of weather snap, matching the symbols to their names.
11 Extreme weather	An open-ended activity in which children draw and describe their own extreme weather picture.
12 The seasons	Children colour a dial to link the seasons to the months of the year in the Northern hemisphere.
13 Going round the sun	Children colour and think about diagrams showing the Earth in relation to the sun's rays in winter and summer.
Local areas	
14 Shelter	A modelling activity in which children create a simple house.
15 Houses around the world	Children colour and label drawings of different house types.

Photocopiable resource matrix

Aim	Teaching points
To introduce children to two objects that dominate the sky and which provide the basis for our notion of time.	Be aware that young children sometimes think the sun and the moon follow them around. Ensure understanding.
To identify Earth's position in the solar system as one of eight planets of differing sizes.	Ensure the pupils understand that some planets are much larger than others and that some, like Mars and Neptune, have a distinctive colour.
To help children understand that day and night are caused by the spinning of the Earth.	The globe is shown tilted at an angle to the sun's rays as it is in reality.
To illustrate that water covers large parts of the Earth's surface.	Ensure that pupils can distinguish between land and water before they start this activity.
To show that plants need water and sunlight to grow.	When they add the arrows, pupils will be creating a simple flow line diagram illustrating a sequence.
To build up children's understanding of landscape features.	Depending on their ability children could either write words to describe the picture or discuss them amongst themselves.
To help pupils understand the basic structure of a volcano.	Pupils will need scissors and glue for this activity – for the best results, duplicate the resource sheet on thin card.
To introduce the idea that we live on a remarkable planet of great beauty and variety.	Pupils could either draw their own pictures directly from the Pupil Book or research their own examples.
To consider how the weather changes and the way that it affects us.	This theme links well with picture book stories and could lead to a large class weather display.
To introduce pupils to symbols in a way that they can relate to and understand.	Pupils will need to cut out the cards and will need to be organised into small groups to play the game.
To explore dramatic and powerful weather events in a non-threatening way.	As well as drawing a picture, children are asked to do some simple writing in this activity. Different levels of support may be required.
To establish the sequence of the seasons using a graphical device.	For simplicity the resource sheet focuses on the Northern hemisphere. Discuss the note at the bottom of the sheet with pupils if appropriate. Some children may have difficulty reading the names of the seasons and months of the year so will require support.
To illustrate how the tilt of the Earth's axis causes seasons to change.	Children need to understand that the sun rises and falls with the seasons; however, understanding the causes can be revisited as they become older.
To generate models which can be used in a class display.	Make an example model to show the class. Consider enlarging the resource sheet to A3 size if possible.
To compare and contrast different house designs.	You could extend this activity by setting up a display of photographs of different houses from magazines or printed pictures from online.

Photocopiable resource matrix

Photocopiable resource	Description
Local areas *continued*	
16 Living in a village	Children annotate a drawing of a small village.
17 Exploring local streets	A survey sheet to help children record findings from a local street work study.
18 Under your feet	Children construct a simple model of a street showing pipes and wires under the roadway.
Maps and plans	
19 Maps and stories	Children complete a drawing showing the route taken by the hare and the tortoise in the story in the Pupil Book.
20 Treasure island	An exercise in which children identify the grid squares of different features on the map in the Pupil Book.
21 Different plans	Children associate drawings of buildings with their plan shapes.
22 The view from above	Children create their own map of a real or imagined place using grid squares.
The United Kingdom	
23 Countries and capitals in the United Kingdom	Children colour the map and flag of the United Kingdom and add labels for the countries and capital cities.
24 Mountains, rivers and seas in the United Kingdom.	Children colour a map showing landscape features of the United Kingdom and turn it into a jigsaw puzzle.
Different environments	
25 Living in the Arctic	Children colour drawings of Arctic plants and animals and add them to a larger picture.
26 Living in the rainforest	Children colour drawings of rainforest creatures and add them to a larger picture.
27 Living in the desert	Children colour and create drawings for a desert mobile.
28 Animals around the world	Children join the dots to create pictures of creatures and then fill in an animal crossword.
World maps	
29 World continents and oceans	Children colour the continents on a world map using the colours specified in a key.
30 World countries	Working from the Pupil Book, children colour flags of different countries.

Photocopiable resource matrix

Aim	Teaching points
To explore the range of facilities that can be found in a village.	Ensure that children identify the shop before they begin.
To focus attention on the range of services which cater for people's needs.	The drawings on the sheet will not exactly match the examples that you find locally.
To illustrate in three dimensions the range of services which reach us by travelling under the ground.	Explain to the children that some of the drawings appear upside down as the model needs to be folded.
To draw attention to landmarks and the sequence in which they appear.	Pupils could add to the 'spectators' by making drawings of other animals.
To develop understanding of alphanumeric grids.	Ensure that the children understand that the letter comes first and the number comes second in a grid reference.
To introduce the notion of plan perspectives at a scale pupils can comprehend.	Making plans of toys is an excellent way both to introduce and reinforce this activity.
To practise using maps and to show how grid squares can support map making.	Discuss with pupils what they want to show on their maps so that they plan what they are doing before they start.
To show the four countries and capitals of the UK.	The children could see how the maps they have created compare with atlas maps.
To consolidate knowledge and understanding of the UK landscape.	It will help if the pupils have coloured round the coast in blue before they create the jigsaw.
To introduce pupils to the distinct flora and fauna of the Arctic.	Pupils will need to draw an Arctic scene before they start to add their animals and plants.
To explore some of the unique features of the rainforest habitat.	Pupils will need to draw a rainforest scene before they start to add their animals.
To focus attention on desert plants and creatures.	Talk about how different plants and creatures manage to live in the desert climate.
To draw attention to notable species which live in different parts of the world.	The children might like to make their own dot-to-dot drawings of other creatures.
To show the different continents on a world map.	Ensure that the children understand that islands form part of some continents, especially in Southeast Asia.
To help children understand that there are lots of different countries in the world.	Ensure that the children look carefully at the flags they are copying rather than guessing the design and colours.

1 Earth, sun and moon

Name ....

1. Colour the picture of the Earth, sun and moon.
2. Add labels to your drawing.

m _ _ _

E _ _ _ _ _

s _ _ _

2 The planets

1. Colour the boxes to make a key. Use a different colour for each planet.
2. Use the key to colour the planets in the picture.

Key

Planet	Colour	Planet	Colour
1 Mercury	pink	5 Jupiter	brown
2 Venus	orange	6 Saturn	yellow
3 Earth	green	7 Uranus	blue
4 Mars	red	8 Neptune	black

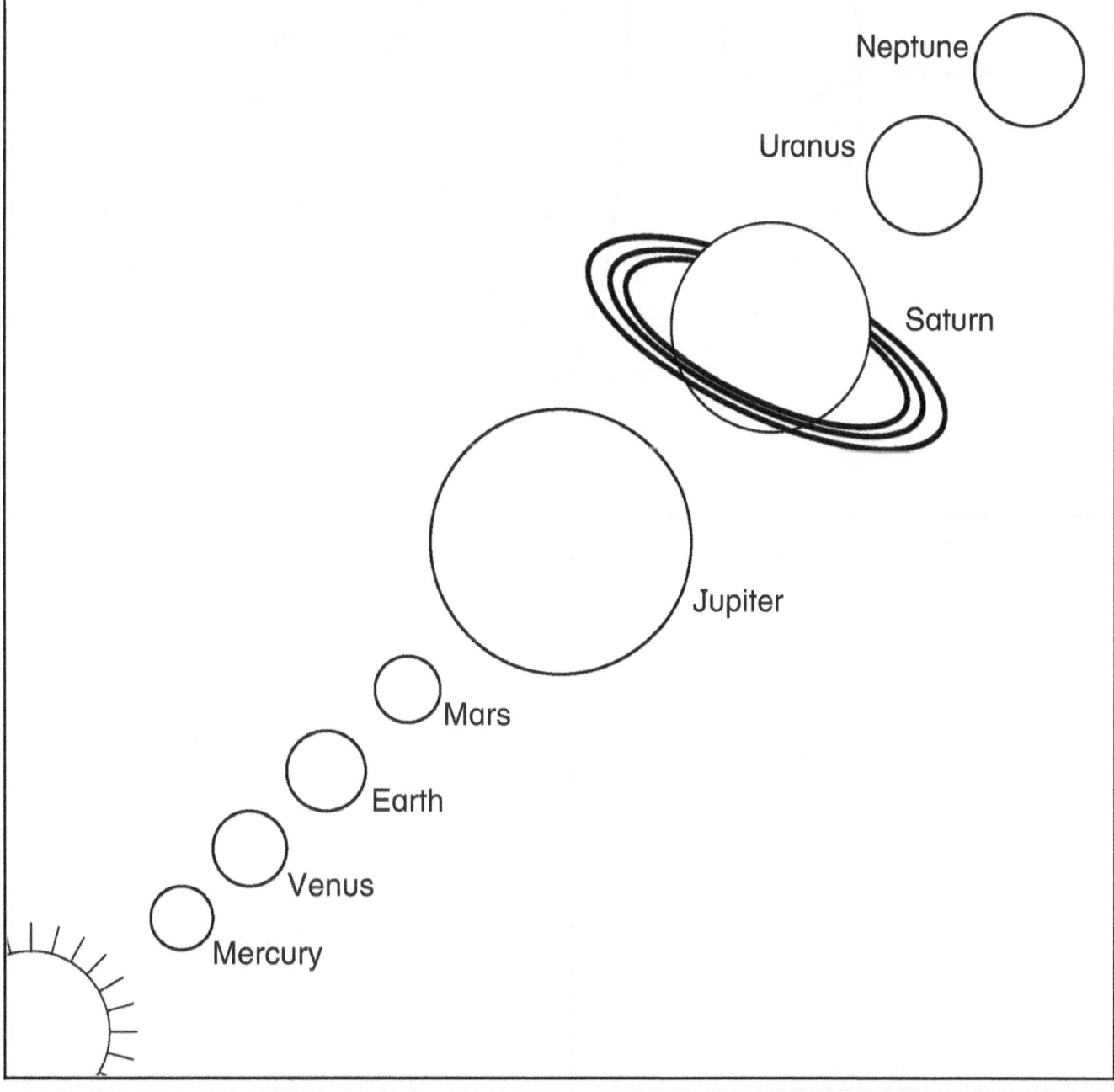

3 Day and night

Name ....

1. Colour the globe to show day and night.
2. Draw a daytime picture and a night-time picture.

North Pole

equator

South Pole

Daytime picture	Night-time picture

© HarperCollins*Publishers* Ltd 2025

Collins Primary Geography
Pupil Book 2: Earth in space pp8–9

4 Land and water

Name ....

1. Colour the boxes in the key.
2. Use the key to colour the globe in the picture.

Key

land	green
water	blue

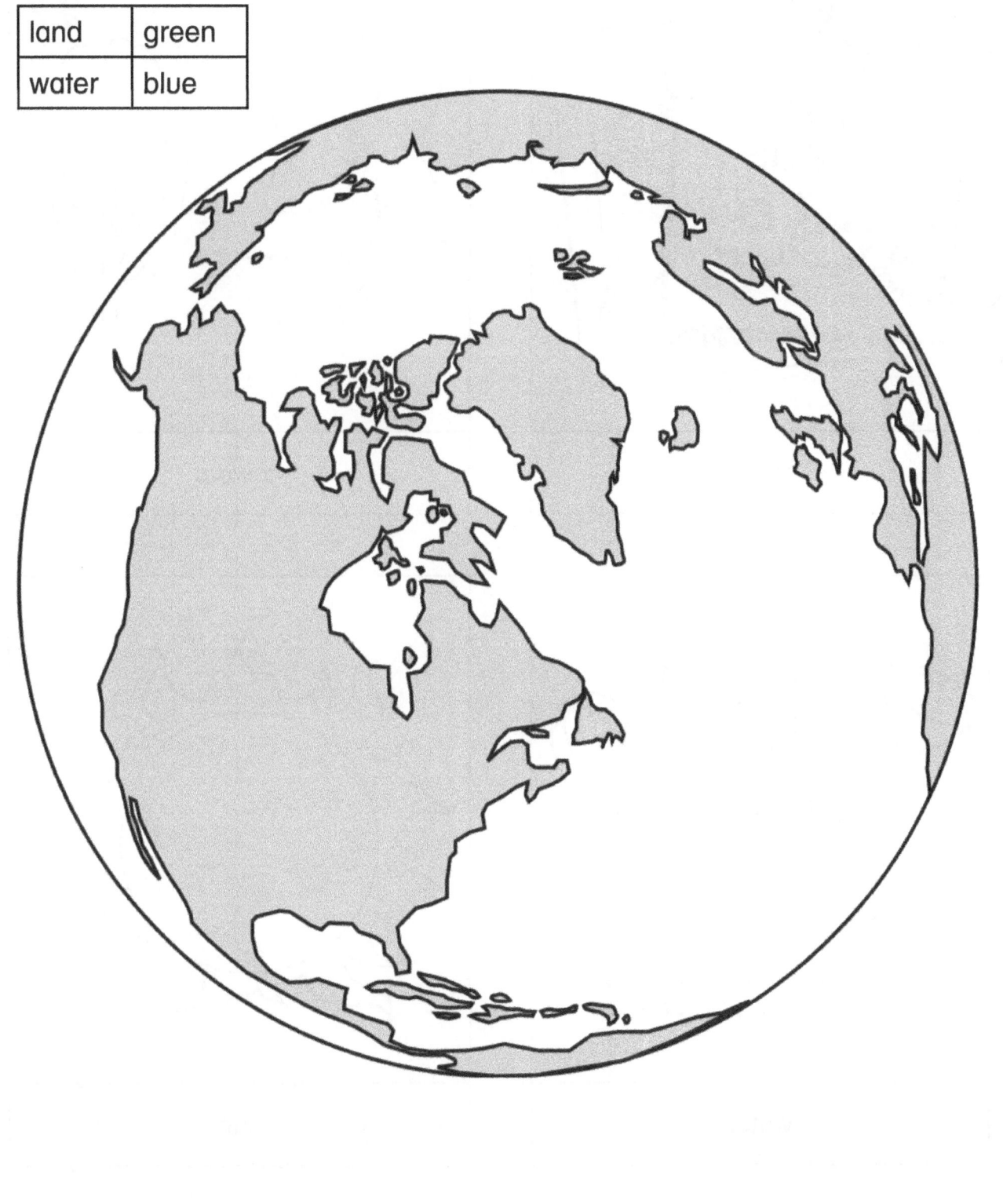

5 A living planet

Name ....

1. Colour the drawings.
2. Draw arrows to link the drawings in the correct order.

seeds

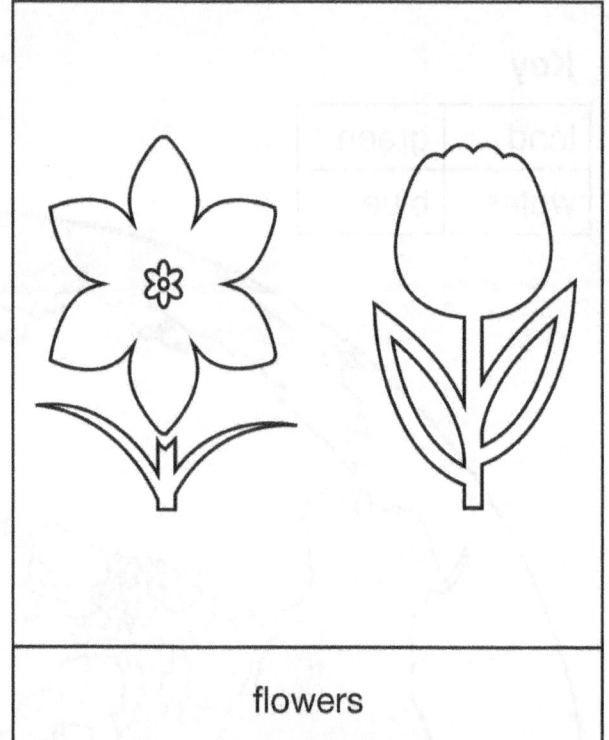
flowers

water

sun

6 The shape of the land

Name ....

1. Colour the drawing.
2. Cut along the lines. Muddle up the squares.
3. Put the picture together again.

Collins Primary Geography
Pupil Book 2: Planet Earth pp14–15

7 Volcanoes

Name ....

1. Colour the model.
2. Cut round the shape. Fold along the dotted line.
3. Stick down the flap to make a cone.
4. Add 'smoke' using tissue paper or cotton wool.

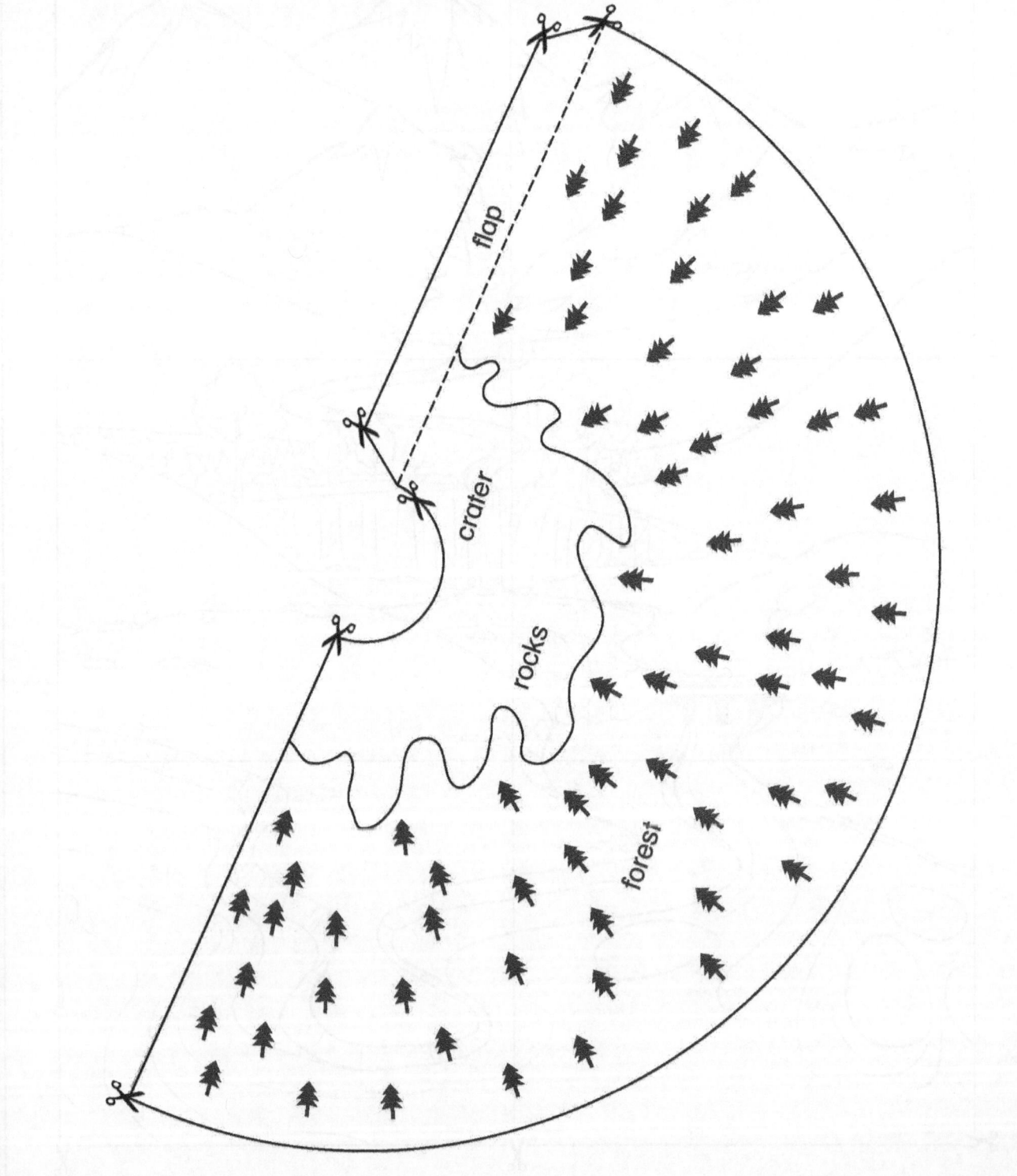

8 World wonders   Name ....

1. Colour the pictures of the different wonders.
2. Draw your own world wonder in the empty box.

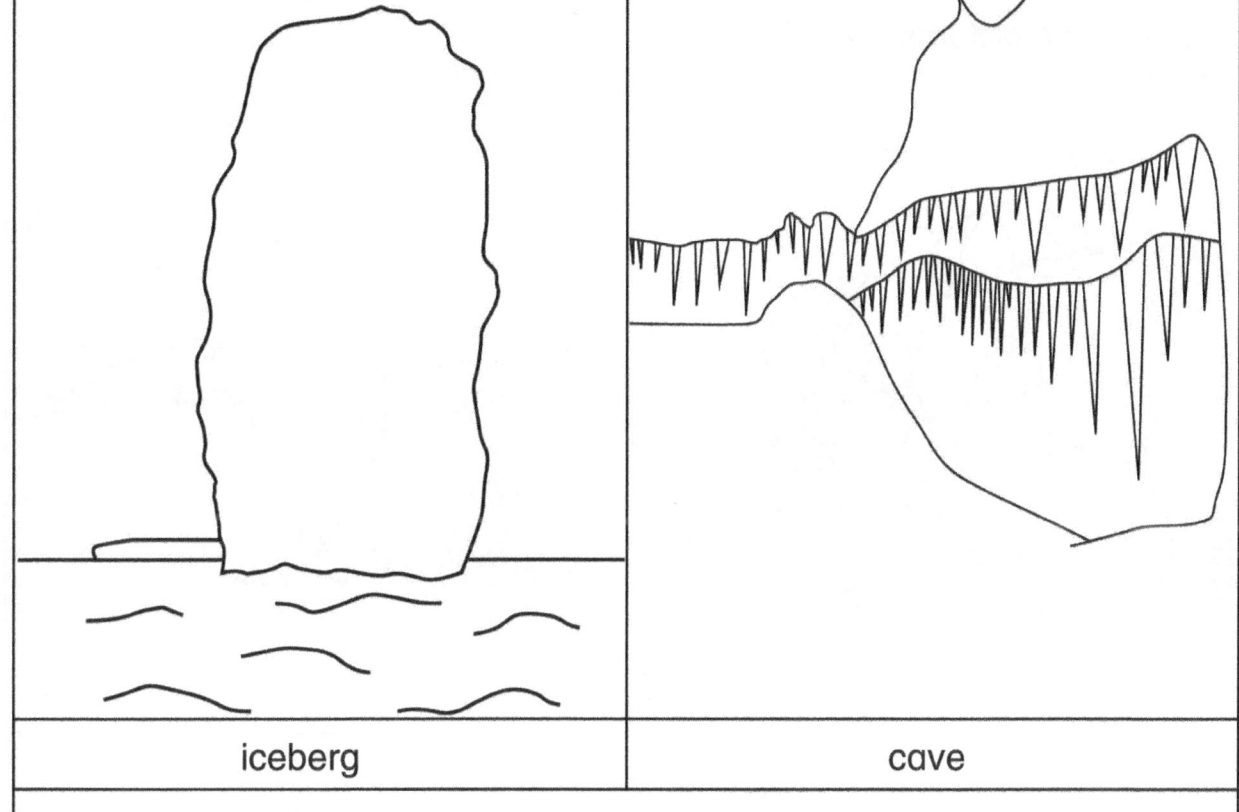

| iceberg | cave |

9. Experiencing the weather

Name ....

1. Label the pictures with these weather words.

   | hot   cold   wet   windy |

2. Colour the pictures.

10 Different types of weather

Name ....

1. Colour the pictures.
2. Cut out the words and pictures.
3. Play a game of weather snap with a group of other children. Match the words with the pictures.

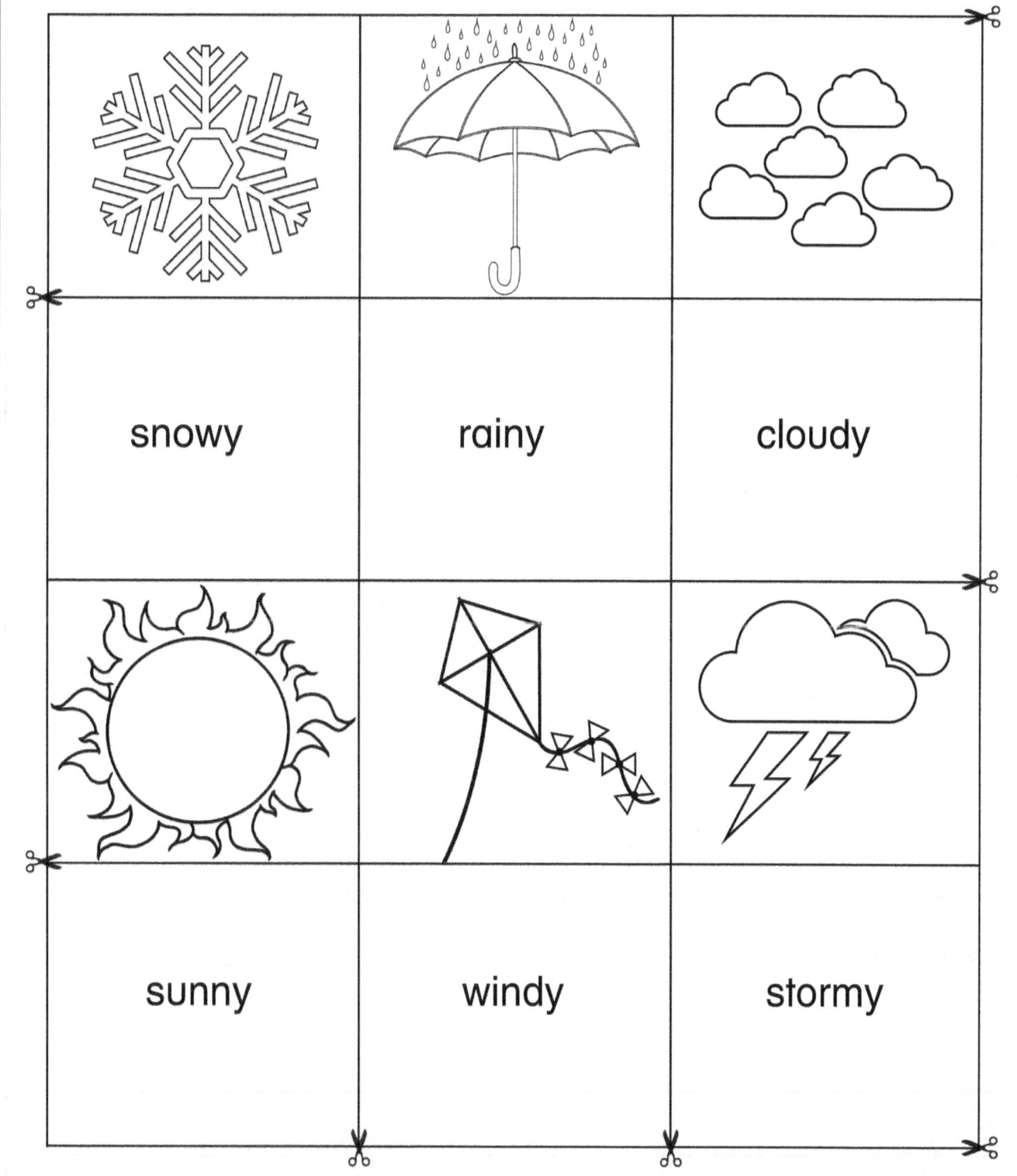

11 Extreme weather

Name ....

1. Draw your own extreme weather picture in the space below.

2. Write a few words to describe what your picture shows.

My picture shows _____

_____

_____

_____

12 The seasons   Name ....

1. Write the name of the United Kingdom seasons round the edge of the dial.
2. Colour the dial and weather symbols.

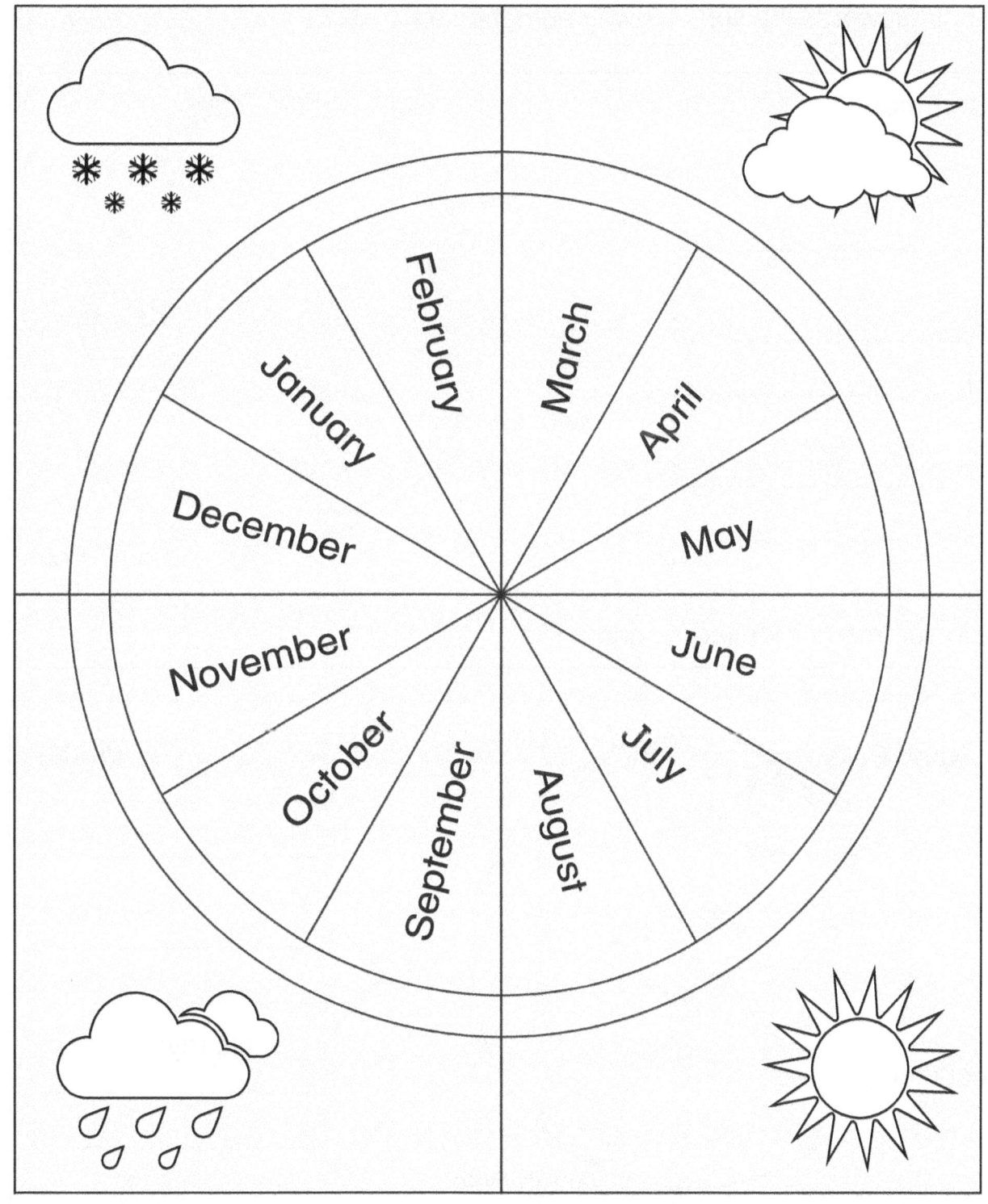

13 Going round the sun

Name ....

1. Colour the pictures.
2. Finish the sentences using the words from the box.

towards the sun   away from the sun

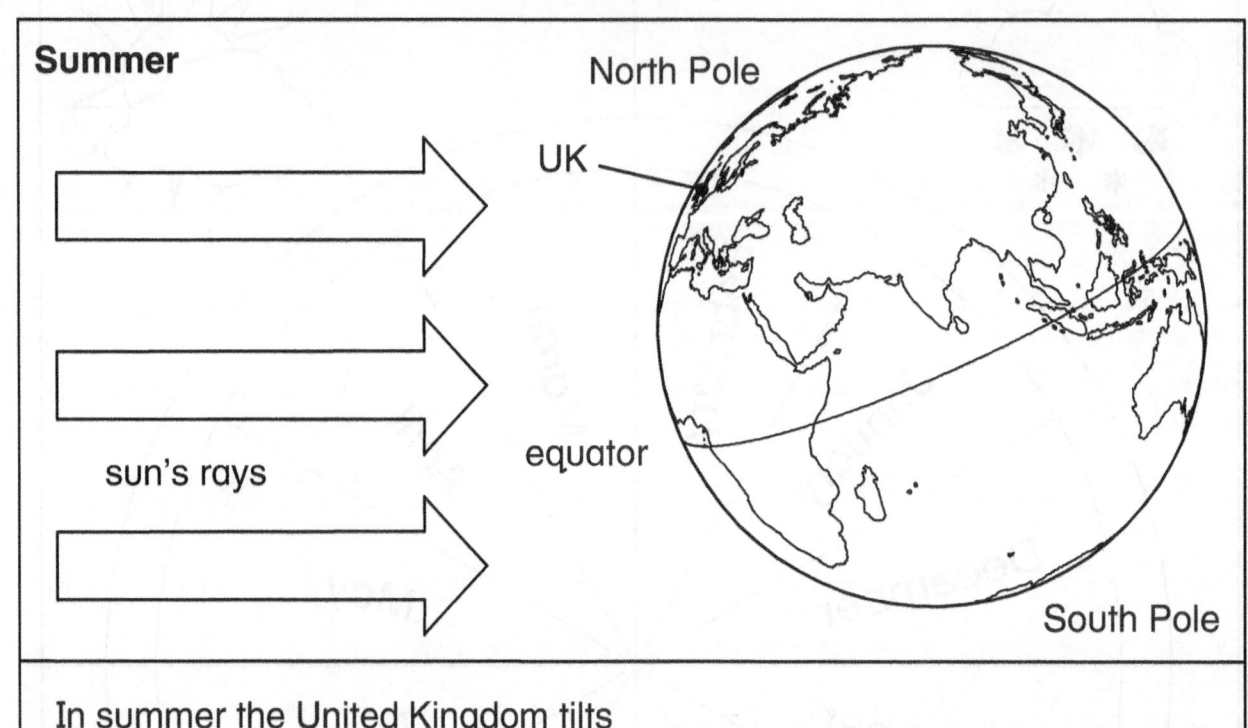

In summer the United Kingdom tilts _____

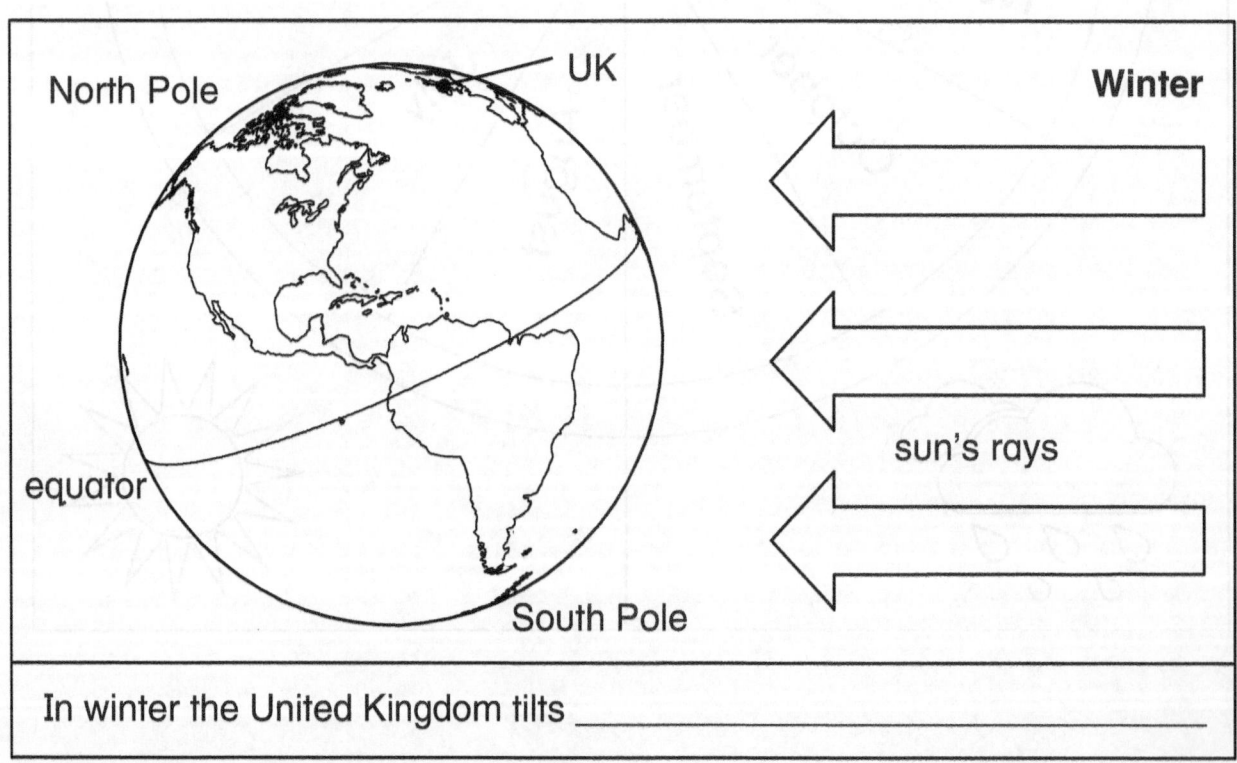

In winter the United Kingdom tilts _____

14 Shelter

Name ....

1. Colour the house cut-out.
2. Cut around the outside and along the lines with scissors. Do **not** cut along the dotted lines.
3. Fold along the dotted lines.
4. Glue the model together.

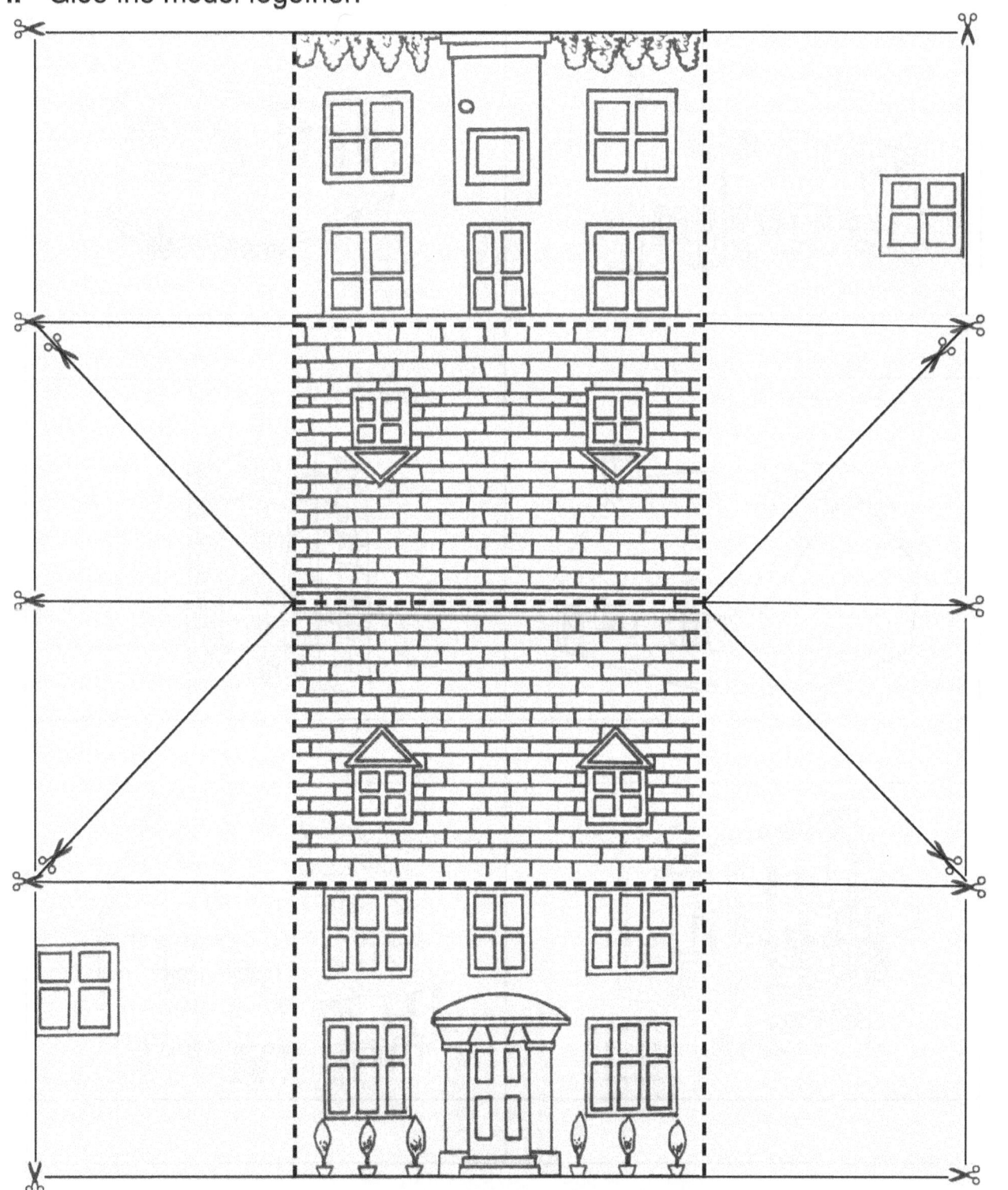

15 Houses around the world

Name ....

1. Colour the pictures.
2. Use the words in the box to label the pictures.

caravan	flats	cottage	stilt-house	terrace	chalet

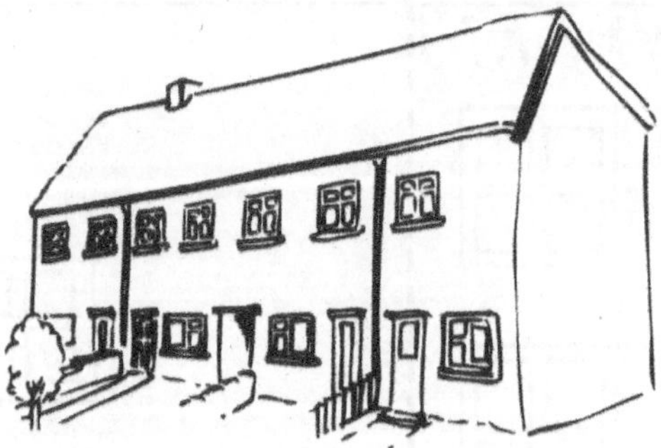

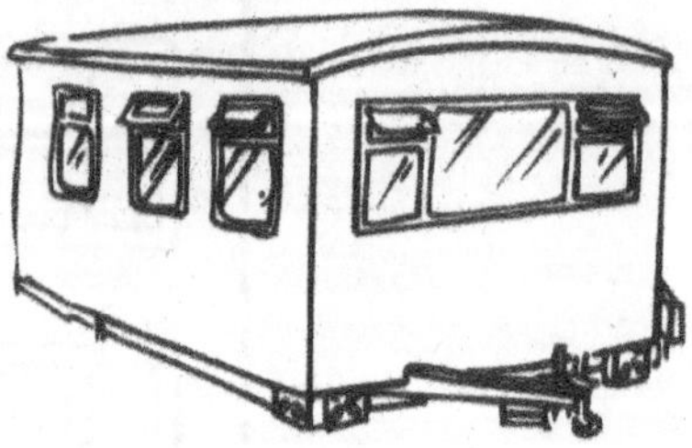

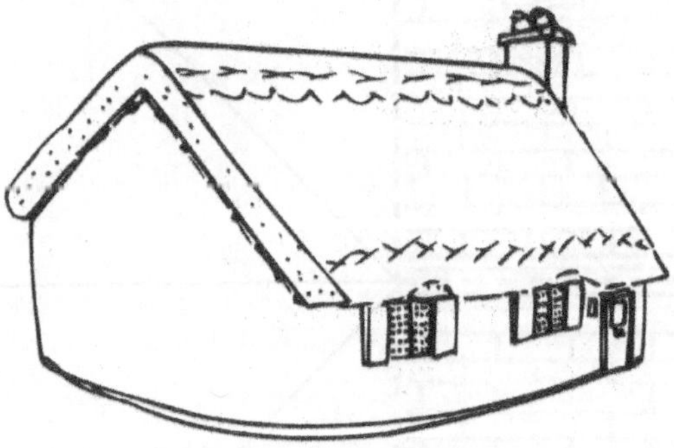

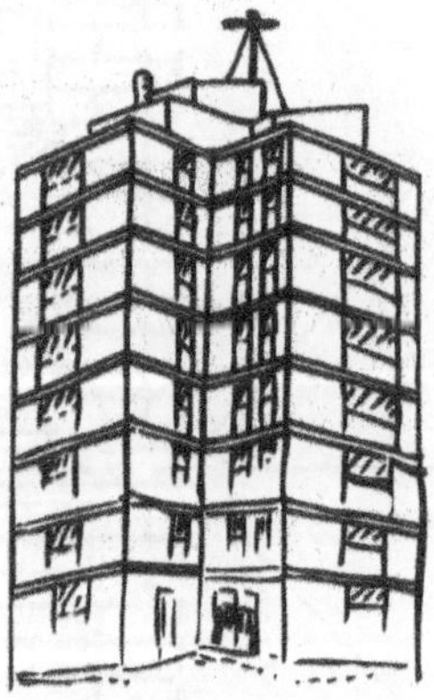

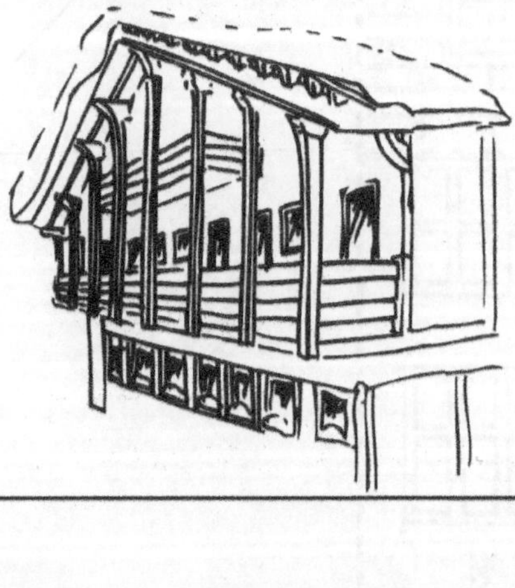

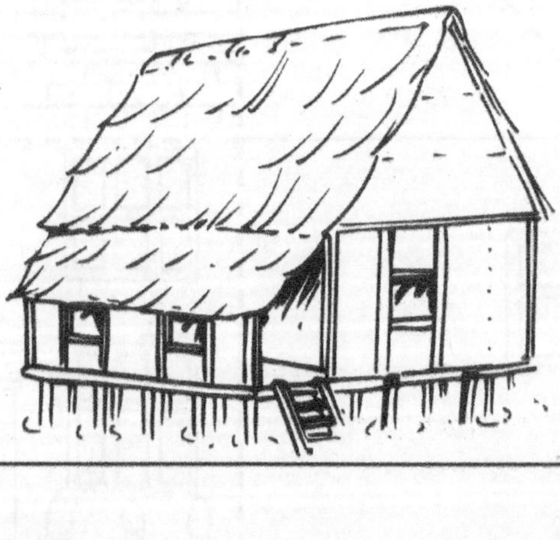

16 Living in a village

Name ....

1. Draw lines from the words to the correct place on the picture of the village.
2. Colour the picture.

farm   fields   houses

church   shop   garage   football pitch

17 Exploring local streets

Name ....

1. Go on a local walk. Tick each of these things as you find them.

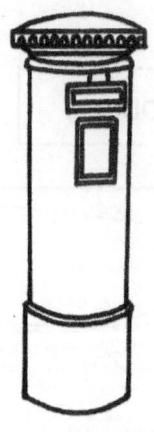

metal covers ☐

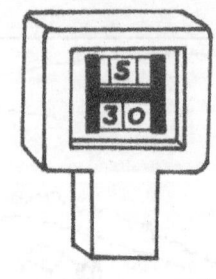

post boxes ☐

fire hydrants ☐

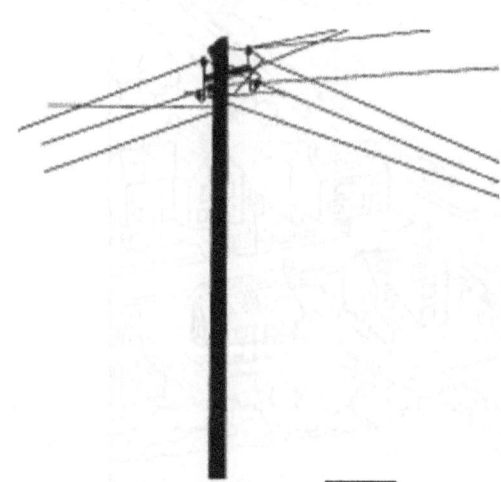

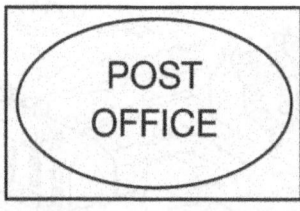

shop signs ☐

overhead wires ☐

advertisements ☐

drains ☐

bus stop ☐

18 Under your feet

Name ....

1. Colour the pictures.
2. Cut round the outside and fold along the dotted lines.
3. Glue down the tab to make a street model.

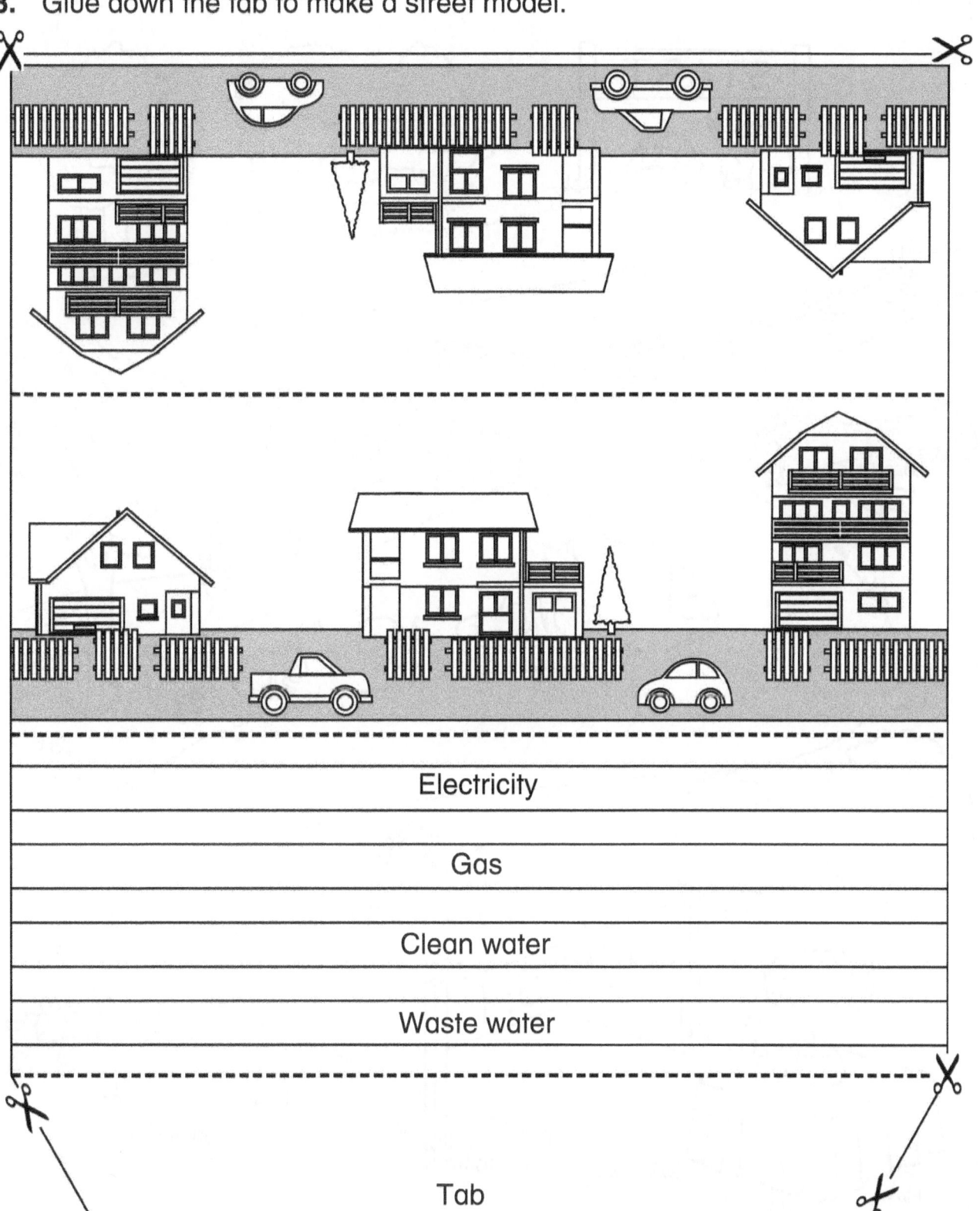

19 Maps and stories

Name ....

1. Colour the picture of the hare and tortoise race.

2. How many creatures are watching the race? ☐

20 Treasure island

Name ....

1. Look at page 43 of your Pupil Book. Write the grid square for the castle and the windmill.

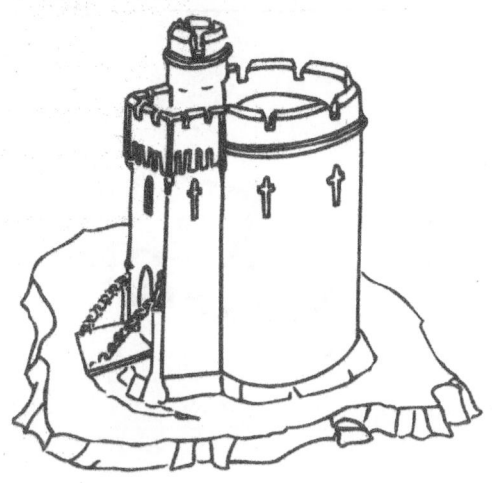 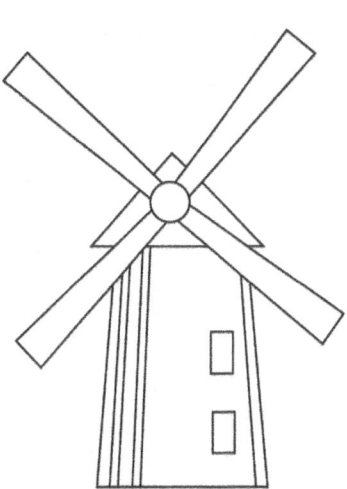

The castle is in grid square ☐

The windmill is in grid square ☐

2. Draw one of the things from these grid squares.

grid square A1	grid square D2

21 Different plans

Name ....

1. Draw a line to link each picture to its matching plan.
2. Colour the pictures and plans.

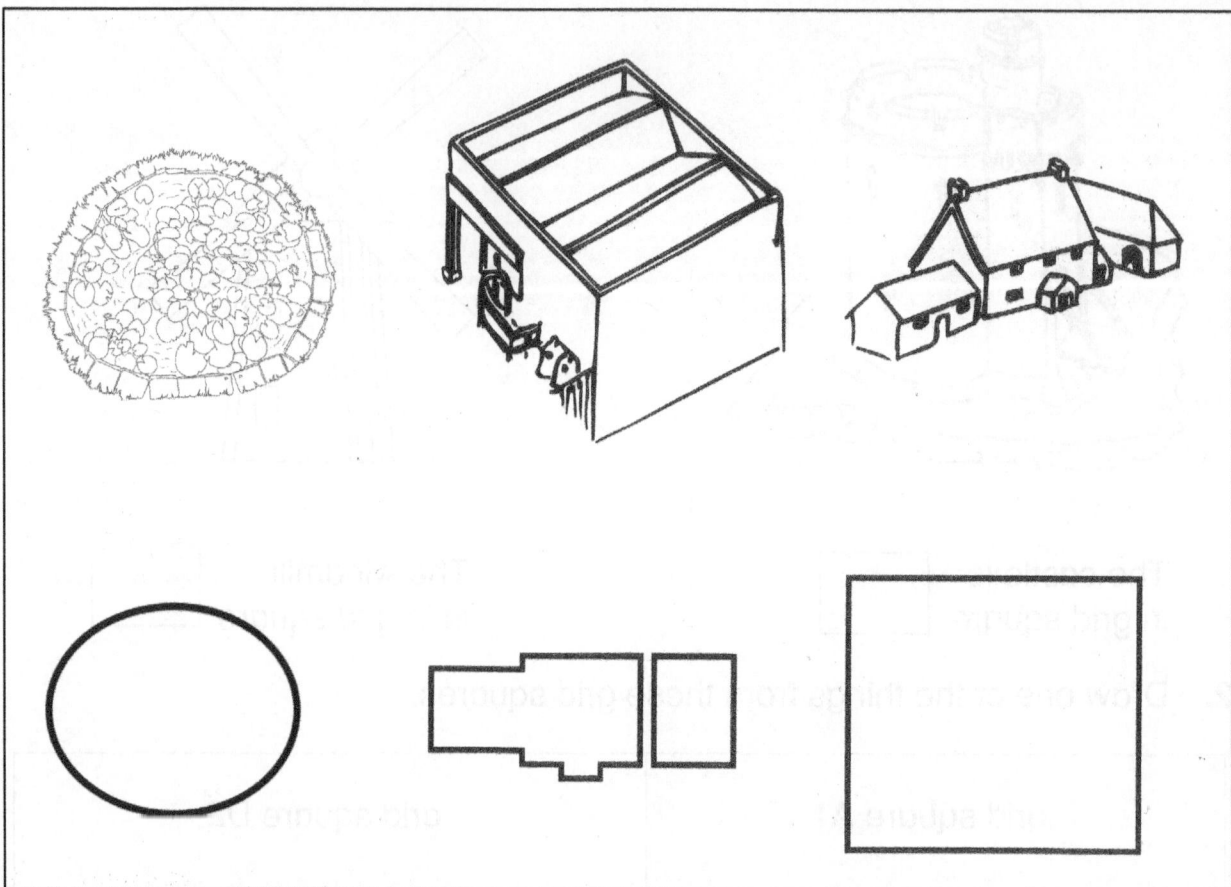

3. Draw a plan of this bungalow.

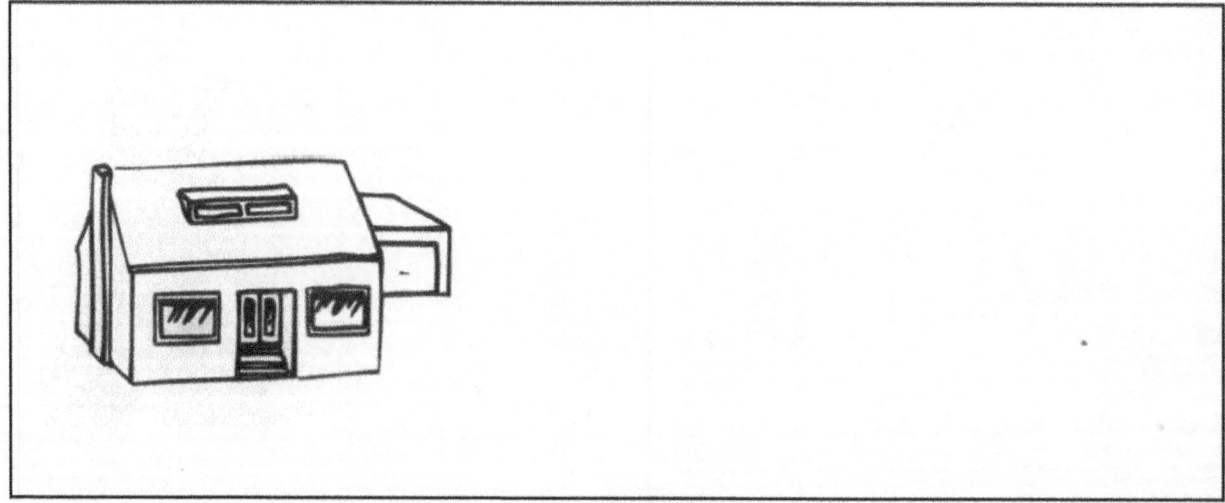

22 The view from above

Name ....

Make your own map looking down from above. You could show buildings, roads, rivers and forests. Add other things as you think of them.

	A	B	C	D
5				
4				
3				
2				
1				

23 Countries and capitals in the United Kingdom

Name ....

1. Colour the map using a different colour for each country.
2. Write the name of each country on the map.
3. Write the name of the capital cities of each country.
4. Colour the flag of the United Kingdom.

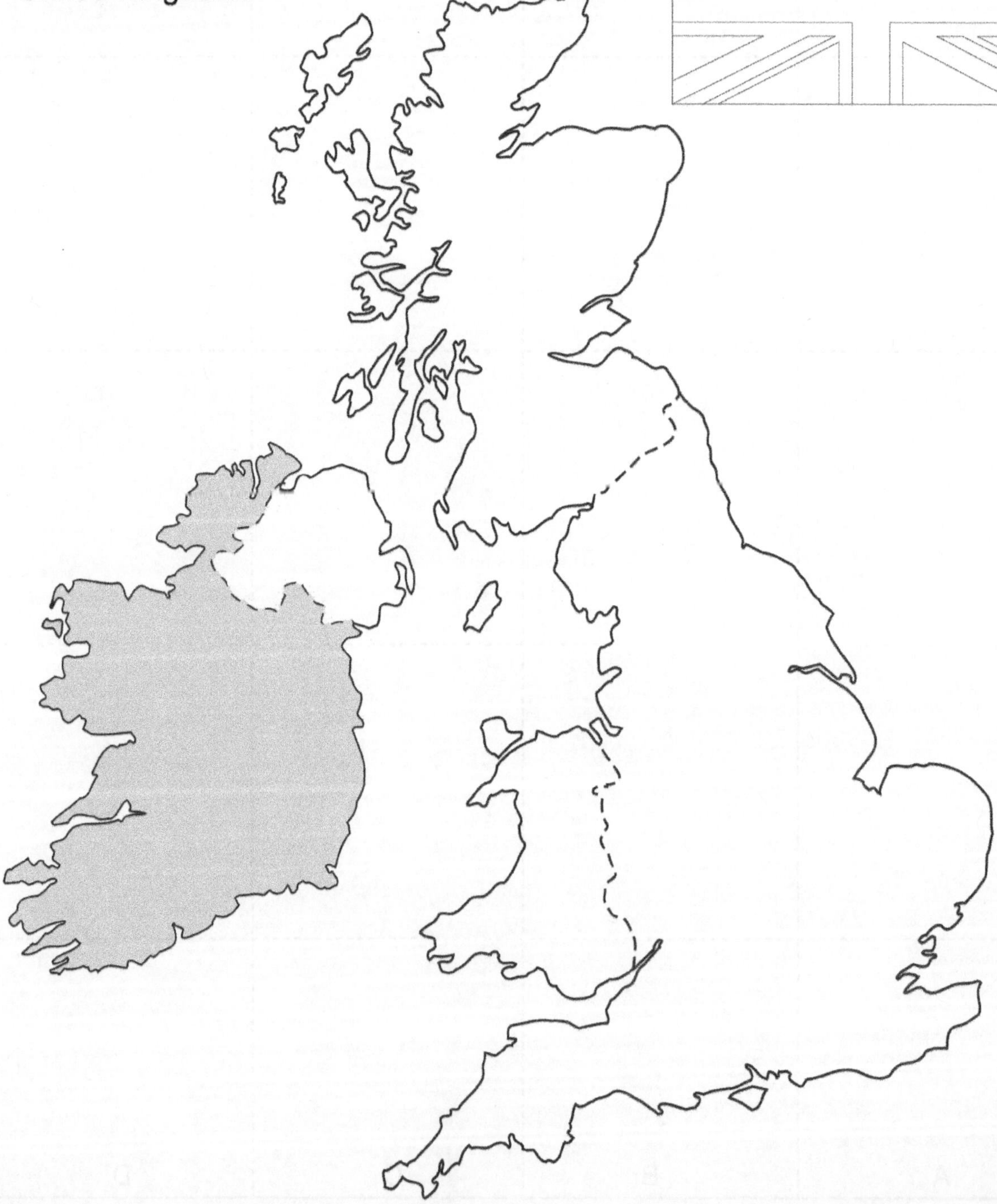

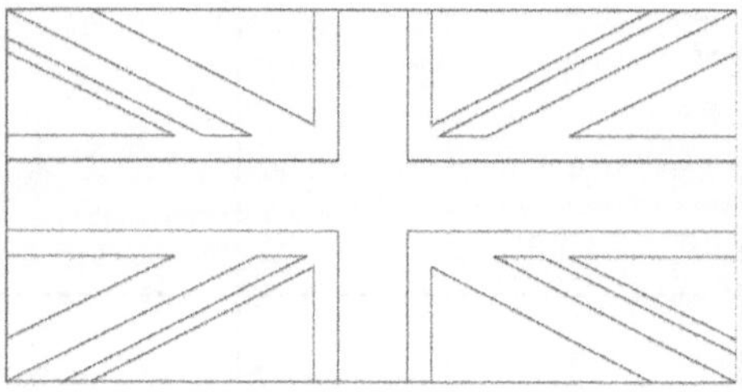

24 Mountains, rivers and seas in the United Kingdom

Name ....

1. Colour the map of the United Kingdom.
2. Cut the map into six pieces, muddle them up and put them together.

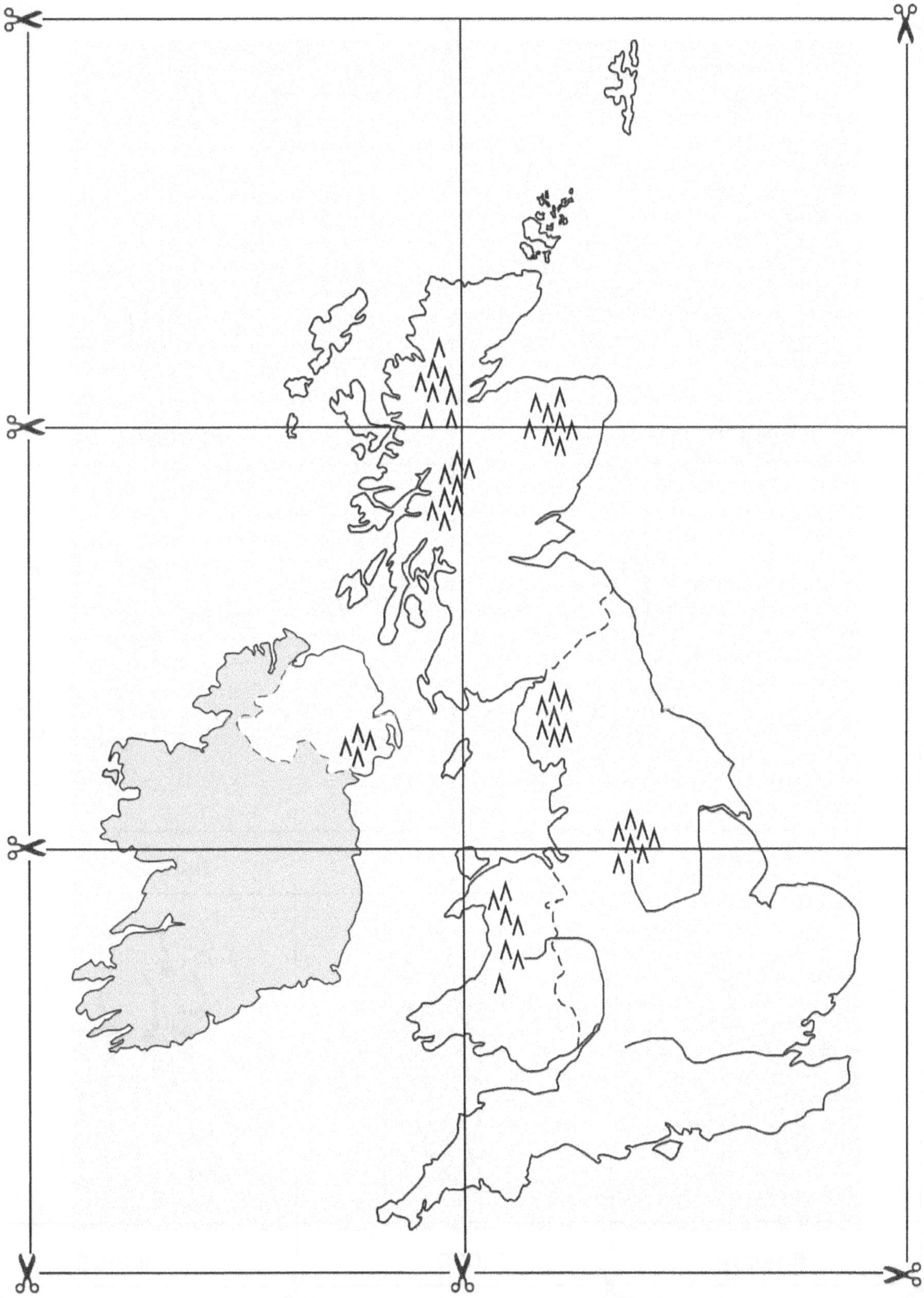

25 Living in the Arctic

Name ....

1. Colour the Arctic plants and animals.
2. Cut them out.
3. Use them on a large Arctic picture of your own.

26 Living in the rainforest

Name ....

1. Colour the pictures of the rainforest creatures. Draw another rainforest plant or creature in the final box.
2. Cut them out.
3. Use them on a large rainforest picture of your own.

27 Living in the desert

Name ....

1. Colour the pictures of the desert creature and plant.
2. Cut them out.
3. Work with other children to make more drawings for a desert mobile.

28 Animals around the world   Name ....

1. Join the dots to make the pictures.
2. Write the names of the animals.

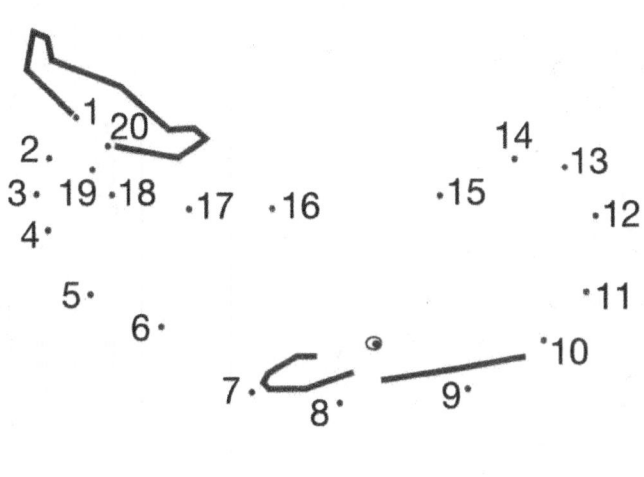
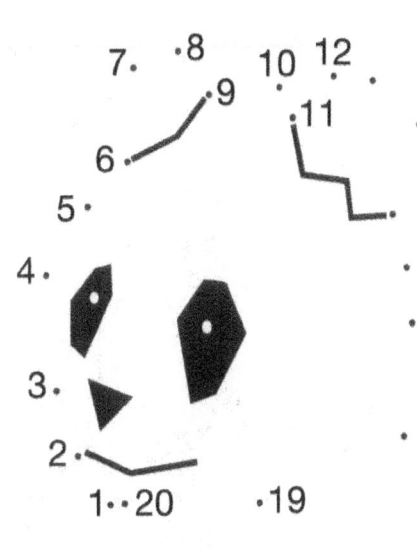

w __ __ __ __   __ __ __ __ __

3. Look at pages 58 and 59 of your Pupil Book.
   Write the names of the animals into this crossword.

b   e

         t   w

      e   p

   p

29 World continents and oceans   Name ....

1. Colour the boxes in the key.
2. Colour the continents on the map to match the key.

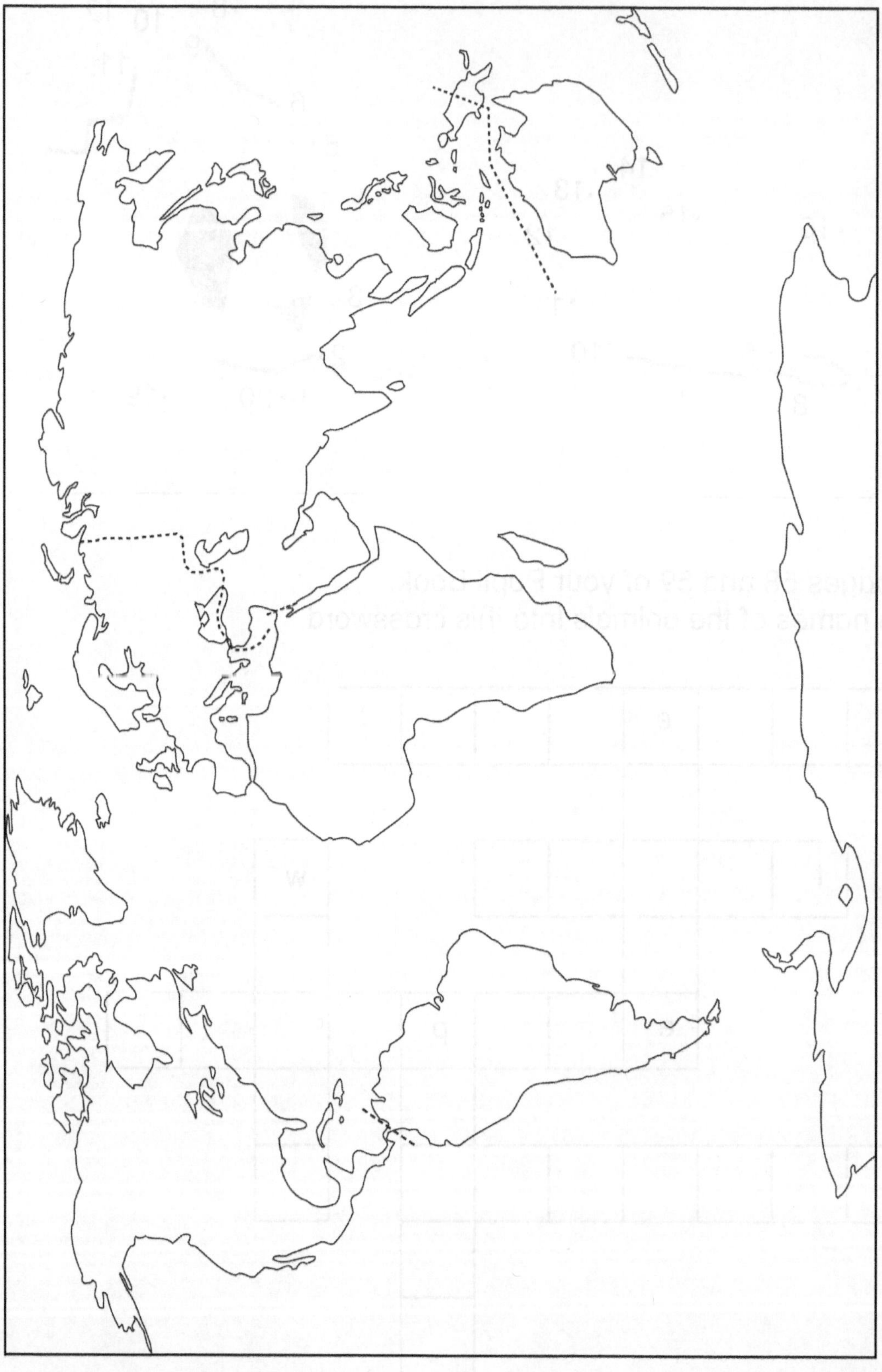

Key	North America	South America	Europe	Africa	Asia	Oceania	Antarctica
	green	brown	blue	red	orange	purple	yellow

30 World countries

Name ....

Colour the flags and write the names of the countries.

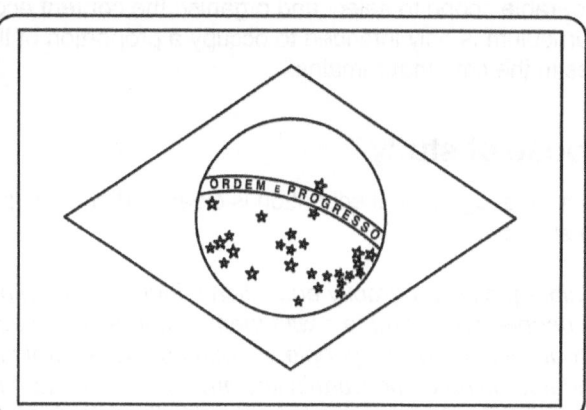

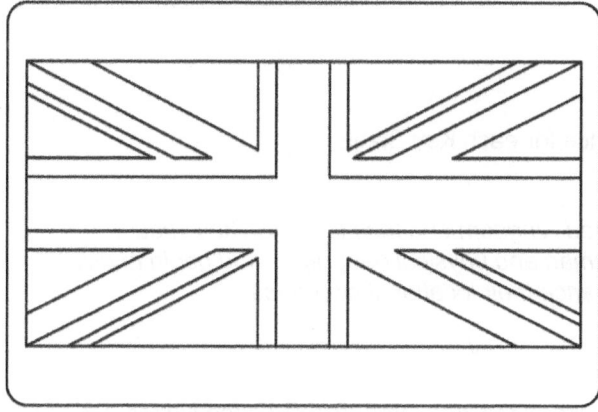

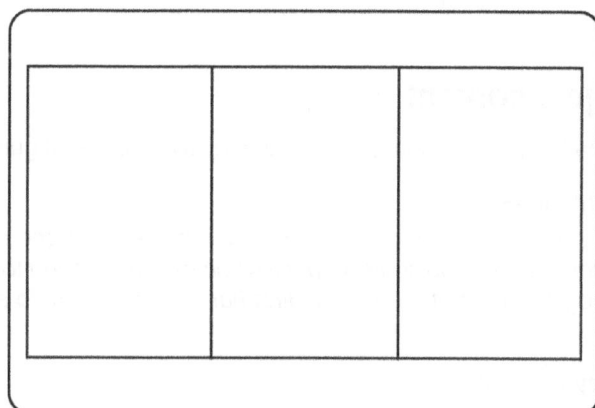

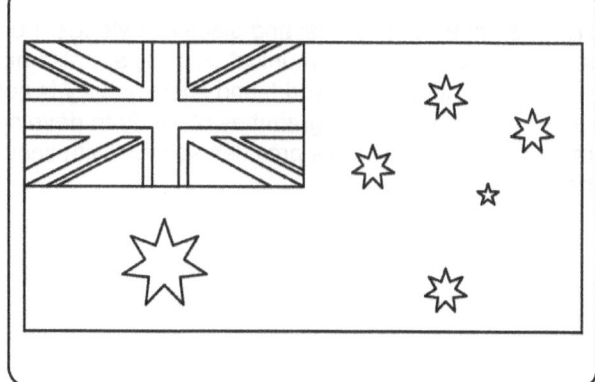

Geography in the National Curriculum in England

The National Curriculum in England provides a geography framework for schools to follow but leaves teachers considerable scope to select and organise the content according to their individual needs. It should also be noted that the curriculum is only intended to occupy a proportion of the school day and that schools are free to devise their own studies in the time that remains.

Purpose of study

The aim of geographical education is clearly articulated in the opening section of the Programme of Study, which states:

A high-quality geography education should inspire in pupils a curiosity and fascination about the world and its people that will remain with them for the rest of their lives. Teaching should equip pupils with knowledge about diverse places, people, resources and natural and human environments, together with a deep understanding of the Earth's key physical and human processes. As pupils progress, their growing knowledge about the world should help them to deepen their understanding of the interaction between physical and human processes, and of the formation and use of landscapes and environments. Geographical knowledge, understanding and skills provide the frameworks and approaches that explain how the Earth's features at different scales are shaped and interconnected and change over time.

Subject content

The National Curriculum provides the following general guidance for each Key Stage:

Key Stage 1
Pupils should develop knowledge about the world, the United Kingdom and their locality. They should understand basic subject-specific vocabulary relating to human and physical geography and begin to use geographical skills, including first-hand observation, to enhance their locational awareness.

Key Stage 2
Pupils should extend their knowledge and understanding beyond the local area to include the United Kingdom and Europe, North and South America. This will include the location and characteristics of a range of the world's most significant human and physical features. They should develop their use of geographical knowledge, understanding and skills to enhance their locational and place knowledge.

There is an emphasis on factual and place knowledge. For example, there is a focus on learning about the UK and Europe. Map reading and communication skills are also highlighted. On the other hand, there are no specific references to the developing world, and sustainability is not mentioned directly. However, there is an expectation that schools will work from the Programmes of Study to develop a broad and balanced curriculum which meets the needs of learners in their locality. This provides schools with scope to enrich the curriculum and rectify any omissions which they may perceive.

Key Stage 1 Programme of study

Key Stage 1 Geography National Curriculum	*Collins Primary Geography* coverage
Develop knowledge about the world	Book 2 (all)
Develop knowledge about the UK	Book 1: Maps and plans; Book 2: The United Kingdom
Develop knowledge about their locality	Book 1 (all)
Locational knowledge	
Name and locate the seven continents	Book 2: World maps
Name and locate the five oceans	Book 2: World maps
Name, locate and identify characteristics of the four countries of the UK	Book 1: Maps and plans; Book 2: The United Kingdom
Name, locate and identify characteristics of the capital cities of the four UK countries	Book 2: The United Kingdom
Name and locate the seas surrounding the UK	Book 2: The United Kingdom
Place knowledge	
Physical and human geography of a small area of the UK	Book 1 (all) *(Book 1 introduces children to the world around them)*; Book 2: Local areas
Physical and human geography of a small area of a contrasting non-European country	
Human and physical geography	
Identify seasonal weather patterns in the UK	Book 2: Weather and seasons
Identify daily weather patterns in the UK	Book 1: Weather; Book 2: Weather and seasons
Identify location of hot areas of the world	Book 2: Different environments
Identify location of cold areas of the world	Book 2: Different environments
Identify the Equator and North and South Poles	Book 2: Earth in space; Different environments
Use basic vocabulary to refer to physical features	Book 1: Planet Earth, Weather, Habitats; Book 2: Earth in space, Planet Earth, Weather and seasons, The United Kingdom, Different environments
Use basic vocabulary to refer to human features	Book 1: Local area, Food and farming, Journeys; Book 2: Local areas, Maps and plans
Geographical skills and fieldwork	
Use world maps, atlases and globes	Book 1: Mapwork skills, Maps and plans; Book 2: Earth in space, World maps
Use simple compass directions	Book 1: Weather; Book 2: The United Kingdom
Use directional and locational language	Book 1: Mapwork skills; Book 2: The United Kingdom
Identify features and routes on a map	Book 1: Mapwork skills, Maps and plans; Book 2: Maps and plans
Use aerial photos and plan perspectives	Book 2: Maps and plans
Devise a simple map	Book 2: Maps and plans
Use and construct basic symbols in a key	Book 1: Mapwork skills; Book 2: Local areas, World maps
Use simple fieldwork and observational skills in their school, its grounds and surroundings	Book 1: (all – Look and find); Book 2: (all – For you to do)

WORLD MAP

WORLD COUNTRIES

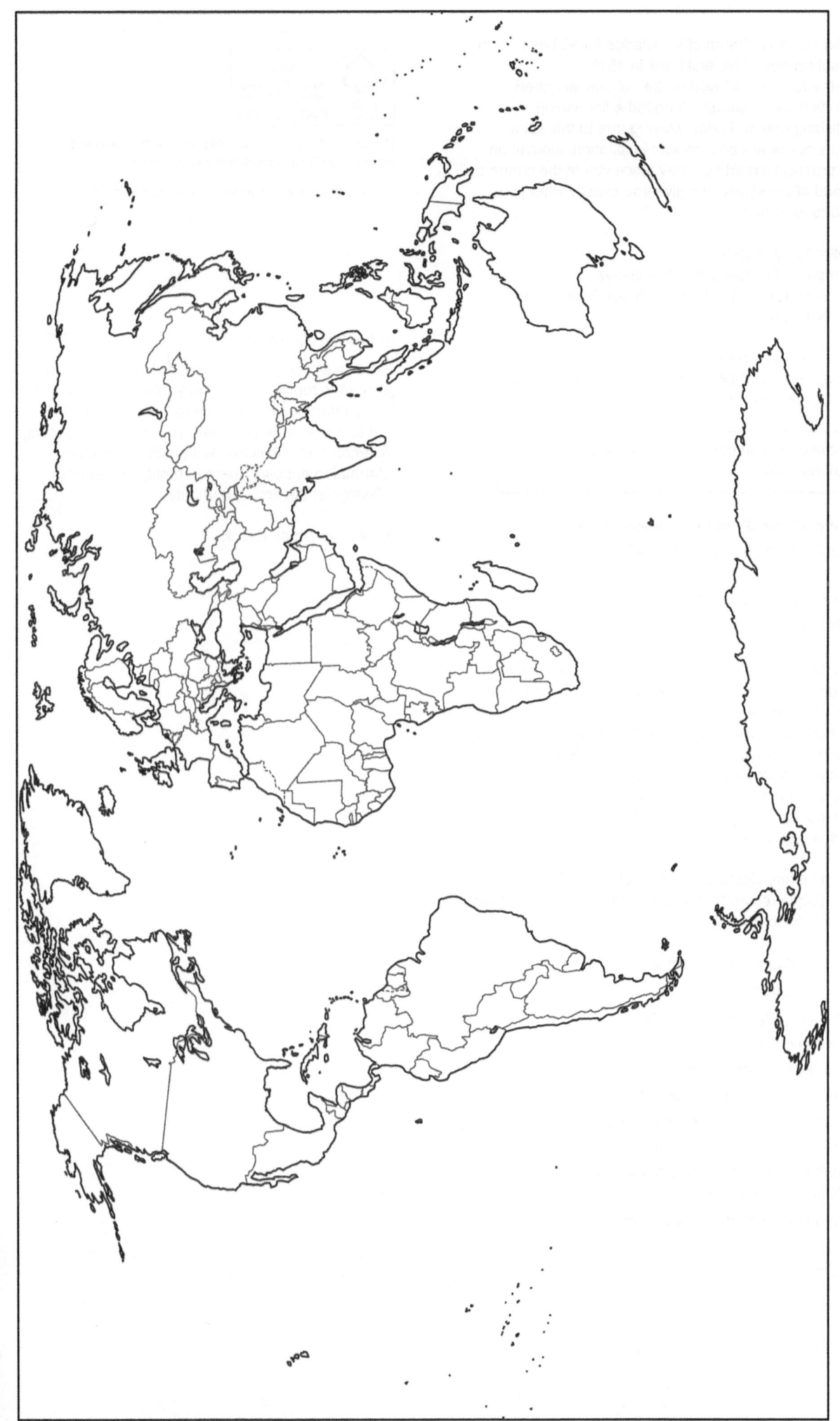

William Collins' dream of knowledge for all began with the publication of his first book in 1819.
A self-educated mill worker, he not only enriched millions of lives, but also founded a flourishing publishing house. Today, staying true to this spirit, Collins books are packed with inspiration, innovation and practical expertise. They place you at the centre of a world of possibility and give you exactly what you need to explore it.

Published by Collins
An imprint of HarperCollins*Publishers*
The News Building, 1 London Bridge Street, London, SE1 9GF, UK

HarperCollins*Publishers*
Macken House, 39/40 Mayor Street Upper, Dublin 1, D01 C9W8, Ireland

Browse the complete Collins catalogue at
collins.co.uk

© HarperCollins*Publishers* Limited 2025
Maps © Collins Bartholomew 2025

10 9 8 7 6 5 4 3 2 1
ISBN 978-0-00-872841-0

All rights reserved. No part of this publication may be reproduced, stored in a retrieval system, or transmitted in any form by any means, electronic, mechanical, photocopying, recording or otherwise, without the prior written permission of the Publisher or a licence permitting restricted copying in the United Kingdom issued by the Copyright Licensing Agency Ltd,
5th Floor, Shackleton House, 4 Battle Bridge Lane, London SE1 2HX.

British Library Cataloguing-in-Publication Data
A catalogue record for this publication is available from the British Library.

Authors: Stephen Scoffham and Colin Bridge
Publisher: Laura White
Product manager: Natasha Paul
Development editor: Judith Walters
Proofreader: Hugh Hillyard-Parker
Cover designer and illustrator: Steve Evans
Internal illustrator: Jouve India Private Ltd
Typesetter: Hugh Hillyard-Parker
Production controllers: Alhady Ali and
   Katie Jean-Baptiste
Printed in the UK at Ashford Colour Ltd

This book contains FSC™ certified paper and other controlled sources to ensure responsible forest management.

For more information visit: www.harpercollins.co.uk/green

Acknowledgements

The publishers gratefully acknowledge the permission granted to reproduce the copyright material in this book. Every effort has been made to trace copyright holders and to obtain their permission for the use of copyright material. The publishers will gladly receive any information enabling them to rectify any error or omission at the first opportunity.

All photos: Shutterstock